ABALONE PIONEERS

The untold stories of the Victorian
Western Zone divers

Rhonda Whitton
and Liz Doran

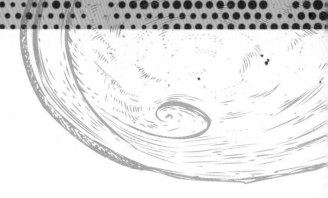

Preface

For the past 16 years I have had the privilege of working with a group of true Aussie pioneers. As young men throughout the 1960s and early 1970s they were attracted to the laid-back lifestyle and potential riches of the budding abalone industry.

In the first years of the fishery there were few rules or regulations, almost no specialised equipment and a poor understanding of the long-term effects of deep diving on the human body. What they did have in abundance was a sense of adventure, ingenuity and a lack of fear.

Using small fibreglass runabouts and cobbled-together dive equipment and lacking maps and depth sounders, they explored the deep waters off the far southwest coast of Victoria. To venture into the deep unknown using this primitive equipment in an area known to be frequented by great white sharks took a level of courage few of us possess.

Over the years many divers suffered the effects of bone necrosis and muscular disorders from their years of deep diving, and a number of deckhands drowned when their small boats succumbed to the power of the sea. Others retired to enjoy the fruits of their labours.

A small hard core have continued their involvement in the fishery up until the present day; they continue to be the backbone of the Western Abalone Divers Association. The common thread that now binds all of the pioneers is an absolute dedication to preserve and enhance the abalone resource that has been such an intrinsic part of their lives. Most of the original pioneers are approaching 80 years of age, so there was a very real need to have their stories documented for posterity.

Today we have a new breed of pioneer divers who are blessed with modern

equipment and technology that the early divers can only marvel at. These divers are pioneering management techniques and data collection that sees them leading the world in cutting-edge fishery management.

This book is a brilliant series of snapshots of individual diver stories. These tales have elements of humour, tragedy, success and failure, but throughout are an inspirational synopsis of the Australian pioneering spirit.

As you read I hope you gain a level of respect, as indeed have I, for the achievements of the unique group of men involved in the abalone industry.

Harry Peeters
Executive Officer
Western Abalone Divers Association

Frank Matthews with a typical catch. *Photo source:* Frank Matthews

A Gelding Street Press book
An Imprint of Rockpool Publishing, Pty Ltd.
PO Box 252
Summer Hill
NSW 2130
Australia

First published in 2019

ISBN 978-1-925946-06-2

Copyright text © Western Abalone Divers Association, 2019
Written by Rhonda Whitton and Liz Doran
Copyright design © Gelding Street Press, 2019

A catalogue record for this
book is available from the
National Library of Australia

NATIONAL
LIBRARY
OF AUSTRALIA

Cover and internal design by Jessica Le, Rockpool Publishing
Layout by Tracy Loughlin, Rockpool Publishing
Cover images by Craig Fox, Tony Jones, Bob Ussher
Glenn Plummer.
Printed and bound in China
10 9 8 7 6 5 4 3 2 1

All enquiries regarding copyright should be addressed to:
Western Abalone Divers Association,
PO Box 5330, North Geelong, Vic 3215

Contents

1. About abalone   1

2. The abalone industry   11

3. Western Abalone Divers Association (WADA)   23

4. The researchers   49

5. Early licence owners   59

6. Other pioneers   81

7. The divers   111

8. The deckhands   167

9. The processors   183

10. The women   197

11. Others reflect on the Zone   203

12. Glossary of terms   221

13. Photo captions and credits   223

1. About abalone

A balone is said to be one of four treasures of the sea, along with sea cucumber (*bêche-de-mer*) and shark's fin. There's debate about the fourth treasure, with some maintaining it is fish maw (the swim bladder of a fish), while others claim it is lobster.

Hardly the prettiest of marine creatures, its rough nondescript shell belies the hidden treasure within. The Chinese refer to the mollusc as 'the grazing cow of the sea' and prize them for their health benefits.

Abalone is a gastropod (stomach-footed) mollusc with an ear-shaped shell. It has a large muscular foot – the edible part – that is protected by an iridescent shell, similar to mother of pearl. It is known as a gastropod because it appears to travel on its stomach. In Australia, abalone is sometimes referred to as 'mutton fish' or 'mutton ear' because its shell resembles an ear. The ear-shaped shell has a row of knobs along the edge and six or seven of these open for respiration.

There are both male and female abalone, but there is no differences in their exteriors. Both are eaten – they are the same quality and size. Approximately one-third of the animal is shell, one-third foot muscle (the meat) and the remaining third offal. The meat is the edible component and the shells are used as decorative items and are a source of mother of pearl.

The exterior of the shell is coloured red, brown and green to blend with their habitat.

The wild waters off Victoria's southwest coast are home to the two main species of abalone taken by Western Zone divers.

Habitat

Haliotis rubra (blacklip abalone) is the main species harvested in Victoria and comprises approximately 80 per cent of the commercial catch in Australia. Adult blacklip grow to 10–20 cm. As its name implies, the side of its foot muscle (the meat) has a black edge (lip). The habitat for blacklip varies: it can be found on reefs, in caves and in narrow crevices. The blacklip prefers comparatively shallow waters around 5 metres but can be found at depths in excess of 30 metres.

Haliotis laevigata (greenlip abalone) are larger than blacklip and grow to more than 22 cm. The shell is rounded, smooth and pale with a chalky texture. This species lives on low reefs and can also be found in rough waters at the bottom of rocky cliffs. It is easy to distinguish from the blacklip because the side of the foot muscle has a bright green edge.

Abalone inhabit vertical rock faces as well as crevices and caves, and cling to the rocky surfaces with their broad muscular foot. These shy creatures are also found under rocks on rocky shores at and below the low-tide line. As light-evading animals, abalone attach themselves to shady parts of rocks with their foot, which has a suction force of more than 4000 times that of their body weight.

The molluscs thrive where the water surges and provides a ready food supply of drifting algae. If the conditions suit they will remain fixed to a particular rock in a single location and will only move to seek out food or during calm weather. They travel by night in search of food, grazing on seagrass leaves and algae growing on rocks. Red algae dominates their diet.

Abalone are prolific spawners and release large quantities of eggs and sperm into the water where fertilisation occurs. However, very few juvenile abalone survive because of their natural predators, which include crabs, starfish, stingrays, some sharks and even rock lobsters.

What's all the fuss about?

Australia produces about one-third of the world's wild abalone, yet it is rarely eaten here because, it seems, the taste and texture don't appeal to the Australian palate. A lot of Australians have tried abalone, usually only once, and most simply shrug their shoulders and ask what all the fuss is about. Some complain that it's tough with a leathery texture; others say it lacks flavour when compared with crayfish, oysters and mussels.

Despite the general lack of appreciation of the mollusc in Australia, it's still acknowledged that there is something exotic about abalone. If you yearn for an unforgettable seafood dining experience and want to taste the mollusc, you'll need to search hard to find it live in seafood tanks in some Asian restaurants. If you can find it on the menu, you'll also need deep pockets because the cost of a meal of wild abalone could be several hundred dollars and even a sliver of the mollusc can send the cost of the dish skyrocketing.

It's a different story in Asia, where abalone is considered a delicacy in

southeast Asian cuisine and is an essential ingredient in many traditional restaurant dishes. Abalone is traditionally eaten for special occasions. The Chinese serve it at wedding banquets as it is said to increase the fertility of the happy couple. And, at the other end of the spectrum, they feed it to the elderly in soups to delay the symptoms of senility.

The quirky television show *Iron Chef* featured abalone as the mystery ingredient in an episode filmed in 1998. The commentator said they'd used US$10,000 worth of abalone during the show. That's HAUTE CUISINE ... to the extreme.

The early abalone divers

Seafood was an integral food source for the traditional inhabitants of the Western Zone, the Gunditjmara and Buandig people, who would have gathered a range of shellfish from these waters, including abalone from tidal pools. When Europeans first settled the area in the early 1830s the settlers saw little value in abalone, either as a food source or a commodity to trade.

Interest in diving for abalone gained some popularity during the 1950s and early 1960s when developments in scuba diving made it possible for non-professionals to dive to depths of 30 metres or more. While the divers developed their skills and some even sold the fish to local restaurants for a bit of pocket money, it was very much a small-scale hobby by a bunch of amateur fishermen.

Fast forward to the hazy, crazy days of the swinging 1960s when abalone divers, like surfers, were a tribe unto themselves. The fledgling industry attracted lots of different people from weird and wonderful backgrounds, with many living fast lifestyles that seemed unsustainable and irresponsible. The idea that a bunch of 'scruffy longhairs' could create an industry based around their desire to play in the ocean doing something they loved was simply beyond belief.

Often these men were fringe-dwellers who took pride in being outrageous — some had a reputation for disgracing themselves by just wanting to have fun. There are probably a few knowing people still around who have a sly chuckle over the goings-on at local pubs where the divers and deckhands held court.

When this first wave of divers moved into the area on their migratory path southwards from Eden in New South Wales or farther south from Mallacoota in Victoria in search of abalone, they were attracted to the abalone industry for its lifestyle. And what a

Clockwise from top left: Tony Jones, Bob Ussher and Noddy Hill with two unidentified men at rear. Noddy Hill. *Photo source*: Bob Ussher. A group of early divers, including Tony Jones (left). Divers on a whale. *Photo source:* Tony Jones

hedonistic lifestyle it was, a lifestyle that was a perfect fit for young fit men who liked being around the water, who didn't much care for a boss looking over their shoulder in a mundane nine-to-five job. Some of them were recreational or spear fishermen. Others spent their days surfing and diving, surfing because it gave them something to do on the days when it was too rough to dive. Some lived on the beach and spent the money they earned on booze and girls. Others lived off the sea, trading abalone and crayfish for essentials.

These hedonistic days were akin to a gold rush. As Len McCall explains, 'We were a bunch of young blokes all eager to get out there and test ourselves with literally no brakes on — and we could make money. I thought I'd do it for a couple of years and then go back to a regular mundane job. Fortunately that didn't happen.'

Certainly the industry attracted its share of flamboyant characters. They were pioneers and, as with pioneers of any sort, not all survived in the industry. Some men found it could be difficult to handle the sudden big money they earned or the trappings that came with it. Many became caught up in the problems the lifestyle attracted — the fast

Top, John Hollingworth and Robert Coffey carting their catch. *Photo source*: John Hollingworth. Wharf shed at Port Fairy. *Photo source* Murray Thiele

pace of living life to excess in all its manifest forms. Suddenly, they had access to the good life, fast cars and women. They were young men with lots of cash, high testosterone and a lot of time on their hands. It was a potent mix.

Establishing an industry

Gary Kenyon, production manager at Sou'west Seafoods, believes the earliest divers sold their abalone to a Chinese man — perhaps Cecil Chang. Cecil travelled through New South Wales into Victoria and then into South Australia buying abalone. Gary believes the man had a processing facility in Mount Gambier and was one of the first to start in the industry back in the very early 1960s.

Frank Matthews, founder of Marine World cannery, says that golfing legend Peter Thomson was the first to realise there was an overseas market for abalone around this time. He recalls how Peter was travelling throughout Asia on a golf tour and became aware of the insatiable demand for abalone there. Peter knew Frank was diving for

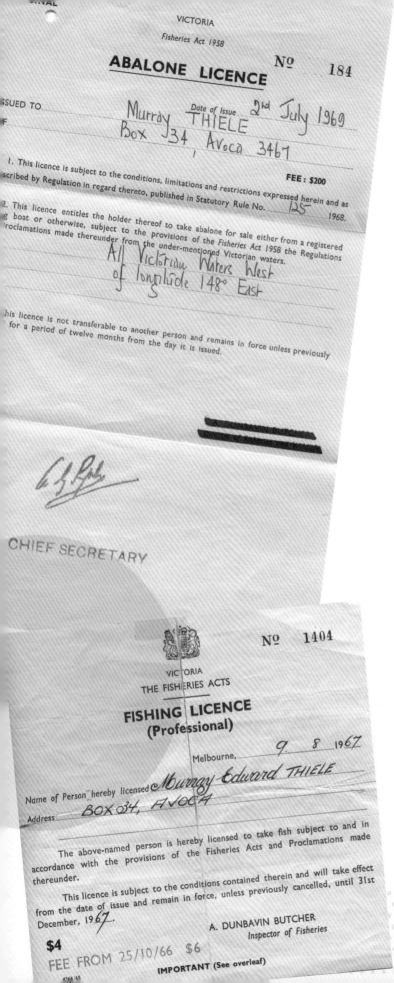

oysters and mussels and he telephoned him to ask whether Australia had abalone. 'I told Peter I knew where there was plenty, so he began searching for markets over there. We set up a little factory and sent product to Singapore. That was the first can of abalone to ever go out of Australia.'

Once the demand was established, in Victoria the focus turned to Mallacoota in East Gippsland when, as Bob Ussher puts it, 'There were abs everywhere.' Many of the divers had moved to Mallacoota from Eden in New South Wales and somehow a newspaper got hold of the story that abalone divers were making huge money. Then according to Bob, 'What could float or anyone who could swim hit Mallacoota.' One day he counted 160 boats going out. They took whatever abalone they could find, from the very small to the large, and within a year the beds had been fished out. In the end they were only getting 30 kilograms a day where previously they'd been getting 700 to 800 kilos.

With an insatiable hunger for the lifestyle that fishing for abalone provided, the men packed their bags and moved south, looking for new and more abundant grounds.

Fraser: We'll protect Aust. fishermen

The Prime Minister, Mr. Fraser, assured fishermen at Portland yesterday that the Government would protect the Australian trawling industry.

He said the primary consideration of the Federal Government would be to protect local fishermen.

Mr Fraser also dispelled rumors that the Government intended to "trade-off" fish in order to again access to Russian beef markets.

"I'm not in favor of any development at the expense of the existing trawling industry in Australia", he said.

Mr Fraser told 21 professional trawler fishermen from Portland, Tasmania, SA, Lakes Entrance and St Remo, that no agreement had been reached with Russia to allow its trawlers to engage in a joint arrangement with local trawlers.

"You've got nothing to worry about," he said.

"I strongly urge no-one to sign anything until we have made a decision.

"Guidelines have to be drawn up for fishing within the 200 mile limit by non-Australians.

"No firm proposals have been put to us yet."

However he did say that overseas countries would be allowed to exploit fishing grounds within the planned 200 mile zone, which were not developed and could not be developed by Australian fishermen within a reasonable time.

Mr Fraser said he could understand the concern by Australian trawler fishermen and said he was concerned himself when he first saw the proposals.

"What the proposals have done is to get a lot of people concerned, and this is good in some ways," he said.

"However it would be unrealistic to

DEAL'S OFF! VINER TELLS

How they did it

In the abalone industry's early days diving technology was primitive, ineffective and in many instances extremely dangerous. The divers worked from beaches with homemade compressors and non-floating hoses. They snorkelled rocky shorelines and used inflated tractor tubes to hold their catch while they fished the inside reefs of enclosed bays. Screwdrivers and levers were used to pry the abalone from the reefs and the men would fill up to six potato sacks with abalone, which were lifted to the surface in buckets. At that time, Frank Matthews recalls, the abalone were as big as dinner plates: 'Wall to wall ab on the bottom. Just mountains of the stuff. Incredible. But it was only worth 25 cents a kilo.'

Frank says the shore crew pulled the diver and his catch by the hose line, hand over hand, through the surf zone and into shore with the catch. They loaded the abalone into plastic garbage bins 'borrowed' from Melbourne's leafier suburbs — that way they were harder to identify. The catches were then manhandled along the beaches to the road, which sometimes meant up high cliffs, where they were loaded onto open flatbed tray trucks for delivery to the processors. All this made diving for abalone an exhausting and risky business.

'Best practice' and 'standards' were terms that were simply not known, or applied, in those early days. A diver's airline was often nothing but a piece of garden hose tied to a compressor with wire. Buckets were used as underwater parachutes to lift the abalone to the surface. Compressor filters were made from anything — most often sanitary napkins.

The unreliable motors the men used as compressors were notorious for playing up in the wet conditions.

It didn't take long before the divers knew they had to improve their working conditions, so they began using boats, bought new compressors and hoses and started using improved diving equipment. This helped them better understand dive medicine and its effects on their body. In the longer term this has made commercial abalone diving a safer occupation than it historically had been.

2. The abalone industry

Abalone is one of Australia's more valuable commercial fishery resources, and the industry is the most highly regulated of the fisheries. Australia produces about 40 per cent of the world's reported wild-stock abalone harvest. Victoria, which produces 1440 tonnes annually, is Australia's second largest abalone-producing state after Tasmania, which produces 2100 tonnes.

The season

Although the abalone season is open all year, if the quota has been filled it automatically closes down until the official start of the next season, which is on 1 April each year. What the divers call a 'shot gun start' often occurs on opening day, as the divers race to be first out. Particularly if the previous season's quota was filled sometime earlier, they want to be quick out of the blocks so they can begin earning money again.

The wild winter weather in southwest Victoria is not conducive to diving, so around mid-year the diving slows down. During spring conditions are variable, and by October and November the weather has improved sufficiently for the men to crank up their diving days so they can catch their quota before Christmas and take the traditional three-month holiday to the end of March.

Licencing

In the earliest days of abalone diving, divers were only required to hold a non-specific fishing licence. Frank Zeigler recalls that Dick Kelly bought his licence for a shilling and grumbled a few years later when it skyrocketed to one pound (approximately $2). Later, fishing licences cost the princely sum of $6.

It is estimated that in 1965 the number of abalone divers had peaked to around 300 and these divers reaped a total harvest of 376 tonnes, live weight. The harvest increased rapidly, peaking at 3384 tonnes during the 1967–68 season. When the fishery was closed to new entrants in 1968, the numbers decreased to 162 divers. The increase in production occurred as many part-time divers left the fishery and those who stayed were relying on abalone diving as their main source of income. At the same time there was a sevenfold increase in annual fishing effort, from an average of about 23 hours per diver to 168 hours per diver.

A $200 abalone fishing licence was introduced as part of the Fisheries Act 1968. This law became effective in May 1970 and further reduced the number of divers to 108. As a limited-entry fishery with non-transferable licences, the number of licenced divers was down to 90 by 1982 through attrition of licence holders. As a result of this limitation of licences production declined steadily to about 2000 tonnes per annum, with some fluctuations upwards, reaching lows of 1143 tonnes in 1977–78 and 1275 tonnes in 1983–84.

Despite the catch rates declining by about 13 per cent between 1968 and 1978, the reduction in annual harvest can be partly explained by reduced fishing effort resulting from diver attrition but mostly by the drop in catch prices during 1977–78. However, decreases in production between 1978 and 1983 occurred concurrently with annual

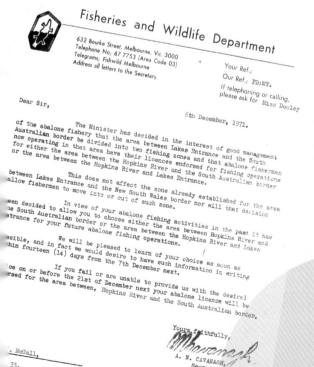

fishing effort increasing to an average of 305 hours per diver in 1982–83 (almost double the average effort per diver of 1967–68) and the lowest recorded average catch rate of 47 kg per hour.

Recent initiatives with respect to the catching and processing sectors of the Victorian abalone industry have involved refinements to the Abalone Quota Management System to provide greater surety of compliance. The main initiatives were the introduction of a sealed bin weighing scheme in 1996, and the implementation of a comprehensive audit trail that covers all phases of harvesting, processing and distribution.

Current catch quotas were initially set on the basis of early assessments and observed stable levels of catch. Victoria's approach has been to adopt a management strategy that attempts to assess how much of the change in abundance at a number of fixed sites in a management zone is due to adjustments in the total allowable catch (TAC). This adaptive management process seeks to determine an ecologically sustainable catch by manipulating the TAC and by monitoring indices of stock abundance.

Zones and catches

In 1968 the Victorian abalone fishery was divided into two management areas: the Eastern and Western zones. In 1970 a third management area, the Central Zone, was added. As part of the zoning arrangement, each licence holder was restricted to fishing in one specific zone and by late 1970 there were 34, 56 and 18 licenced divers in the Eastern, Central and Western zones respectively.

--

Legal minimum lengths (LML)

Legal minimum lengths (LML) were introduced as a strategy for managing the fishery and catch quotas were initially set on the basis of early assessments and levels of catch. For blacklip abalone the LMLs were 10 cm for Port Phillip Bay, 11 cm for all other areas between Lakes Entrance and Lorne, and 12 cm elsewhere. A single LML of 13 cm was set for greenlip abalone in all areas.

Licence consolidation, 1984

During 1984 the Fisheries (Abalone Licences) Act was introduced to permit licence transferability on a two-for-one basis (ie, a diver needed to obtain two existing licences which were then consolidated into one new licence) to encourage new, generally younger divers into the fishery without causing an unsustainable increase in catch. Licence fees were based on the average price per kilogram paid to divers for the preceding year's catch.

However, the entry of highly motivated divers, who had each paid $100,000 to $160,000 to purchase two consolidated licences plus a $10,000 licence transfer fee,

led to a substantial increase in production to 1900 tonnes during 1987–88. A threefold increase in price was paid from an average of $4.59 per kg live weight in 1983–84 to $15 in 1987–88. This provided an additional incentive to increase fishing effort. Some divers were landing about 40 tonnes of abalone annually.

Total allowable catch (TAC)

The abalone fishery introduced TAC in 1988, under the Fisheries (Abalone) Act 1987, to control the size of the catch. Zone boundaries were re-defined to 148° E (between Central and Eastern zones) and 142° 31'E (between Western and Central zones), and separate TACs of 460, 700 and 280 tonnes were set for the Eastern, Central and Western zones respectively. These TACs were equitably distributed by allocating a quota of 20 tonnes per annum to each licenced abalone diver in the Eastern and Western zones, while each Central Zone licence holder was allocated 20.58 tonnes.

Also, licences became transferable on a one-for-one basis. Each diver's quota allocation effectively became an individual transferable quota (ITQ). The total TAC of 1440 tonnes was set to limit the catch to about 70 per cent of the production in the early 1970s. Licences and quota cannot be transferred between zones.

Since the introduction of licence transferability and consolidation, the number of licence holders has been 23 divers in the Eastern Zone, 34 in the Central Zone and 14 in the Western Zone. Similarly, divers' ITQs have not changed since the introduction.

Quotas are reviewed on an annual basis after a consultation process between the industry representatives and Fisheries. What the industry wants they don't always get, and sometimes the decision isn't known until the start of the season on 1 April.

The virus

In 2006 the abalone viral ganglioneuritis (AVG) devastated key fisheries scattered along 200 kilometres of Victoria's southwestern coast. The virus is a highly virulent herpes-like virus that affects the nervous tissue of abalone and causes death in a short space of time. There are no known public health or food safety implications associated with this virus.

After the virus was detected, Western Zone divers fished to a reduced zonal quota of 20 tonnes per year. The zone's abalone divers are confident the industry will bounce back from the disease that cast a shadow over it for almost three years.

Poaching

Although abalone fishing is strictly regulated and heavy fines are enforced for those who break the rules, this hasn't stopped unscrupulous unlicenced fishermen from poaching abalone in the Western Zone.

Konrad Beinssen recalls that while he was conducting research in the Western Zone he came across many areas that had been poached and knew of a couple of

well-known poachers. The poachers took enormous risks to get an illegal catch, such as working Lady Julia Percy Island at night where the waters are notorious for sharks, particularly white pointers.

Konrad describes the notorious poacher Cam Strachan as being a 'nice bloke'. He adds, 'They should have given him a licence in the early days — he was so resentful at having been rejected for a licence. He had an early history in the industry and probably deserved a licence.'

Ron O'Brien believes poachers did real damage to the Zone, first targeting the inshore shallow grounds with shore access. He maintains they also hit Whites, Murrells and Inside Nelson. The most resilient beds for poachers were The Passage, front of Nelson, front of Bridgewater and Whites Beach.

Phil Plummer lives just outside Port Fairy and has seen poachers at work near his home. One day he was taking his dog for a walk along the beach and saw a group of men sitting on rocks. The men didn't look suspicious, he says, but he knew what they were up to when his dog charged at them and returned carrying abalone meat. Phil phoned Fisheries but by the time they arrived the men had disappeared. The poachers weren't too bright, because they got caught when they came back to collect their 30 undersized abalone. Phil described the men as serious poachers, as they took all the abalone irrespective of size.

Dick Cullenward lived on the beachside of the road just outside Port Fairy, and like Phil could see poachers at work from his home. He says he'd see them leave their abalone in a bag on the bottom and return at night to collect it — he would see a flashlight as they came to pick up their catch. Although he reported the poaching activity to the authorities, he believes the penalties had little deterrent effect.

The perils of diving

A weird lot

People living in the southwest corner of Victoria were a notoriously conservative bunch. Fishing was an important industry in the region and locals were used to the idiosyncrasies of fishermen, but abalone divers were considered even more eccentric.

In the early days, the men were diving for

the big money they could earn. They needed the money to sustain their lifestyle and it was well known that they only wanted to be paid in cash. This meant that the local banks had to have tens of thousands of dollars in cash on the days the men were diving and arrange for armed guards to accompany the deliveries. The banks also had to learn about international banking, letters of credit, exchanging matching documents and a host of other banking practices not previously required in the sleepy coastal villages.

The *Portland Observer* of 27 January 1971 offered insightful observations about the local abalone divers and their way of life. 'Four years ago, abalone was a foreign word to most Portland people. Today, the word abalone conjures up visions of divers braving sharks and the bends, small boats in wild seas and a new breed of people making big money from the sea-bed. But, as with all stories, there is another side to this fascinating industry. The abalone industry invaded Portland about three and a half years ago. With the industry came new people, people who were close knit and, in some respects, different to the original townsfolk. Many had long hair, others were highly educated and all lived a free life. The industry appeared exotic and seemed even more so when stories of divers making $1000 in little more than a week began circulating … Legends grew up about the daring of the "ab boys". Why do these people risk their lives for the abalone shell? One of the main answers is money. Another … is because the people involved in the industry like the life.'

The Island

Lady Julia Percy Island, 22 kilometres southwest of Port Fairy, is one of four Australian fur seal breeding colonies in Victoria and is possibly the largest colony in Australia. The Island is off bounds to visitors except on organised tours. It is also legendary among the Zone's divers. The pristine waters surrounding the Island coupled with its remoteness have created an abundant abalone population.

When abalone is in short supply in other areas along the southwest coast, divers know there is always a good catch to be had at the Island. The Island strikes fear into the hearts of many a diver because it is an important hunting ground for great white sharks (aka white pointers), which prowl the waters in search of fur seals.

Diver Henri Bource lost his leg during a dramatic encounter with a great white off Lady Julia Percy Island in 1964. He was with two other divers off the Island when a great white came from under Bource, attacked him and took his leg off at the knee. His diving partners managed to get Bource back on board their boat and then radioed to shore for help. Bource later described how he tried to get his leg free by jamming his hand down the shark's throat and gouging its eyes. A few years later Bource, an amateur underwater photographer and filmmaker, used the film footage of the attack to create the documentary *Savage Shadows*. If you've got a strong stomach, you can see footage of Bource and the shark on YouTube.

Being bent

The bends is an ever-present spectre for abalone divers every time they dive. The bends results from diving too deep for too long and/or not surfacing slowly enough to allow the body to adequately decompress. This causes nitrogen bubbles to form in the body's bone marrow and escape through the bone joints, resulting in damaged tissue and chalky bones. The bubbles can form in, or move to, any part of the body and the effects on the diver range from joint pain and rashes through to paralysis and even death. Pain can be anything from mild to excruciating.

The bends affects all divers differently — some are more susceptible than others. Age, body type, fitness and ambient temperature are all factors that affect a diver's vulnerability to the condition. Pain most frequently occurs in the shoulders, elbows, knees and ankles, while headaches, skin complaints and visual problems are also common.

Breathing 100 per cent oxygen reduces the nitrogen in body tissues and can provide effective protection, but breathing pure oxygen alone does not provide full protection. In severe cases, divers are treated in a decompression chamber, which allows them to undergo their decompression stops in a controlled environment. Portland now has a decompression chamber but before that the Zone's divers had to be taken to either Melbourne or Sale for recompression.

Bob Ussher was quoted in a lengthy undated article published in Adelaide and written by Chris Pash. Bob described how he was diving one day when his compressor failed and he could not breathe: 'You have

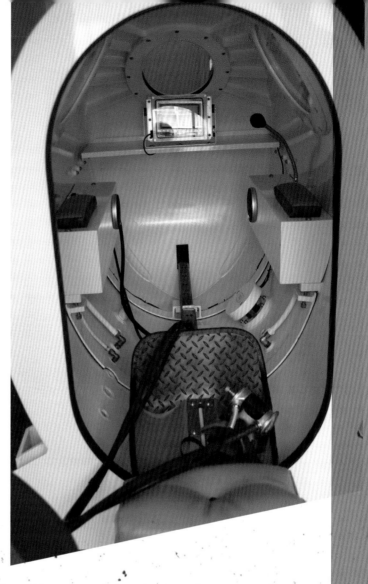

one breath and the next breath it's gone,' he said. He shot to the surface once he shed his weight belt. 'All I was thinking of was getting to the surface. I didn't know what was going on. I knew I was coming up far too quick and tried to have a duck dive to slow myself up. The next thing I knew I was on the surface and climbed on board the boat. After about five minutes my shoulder started to pain and I thought it was just a normal little niggle of

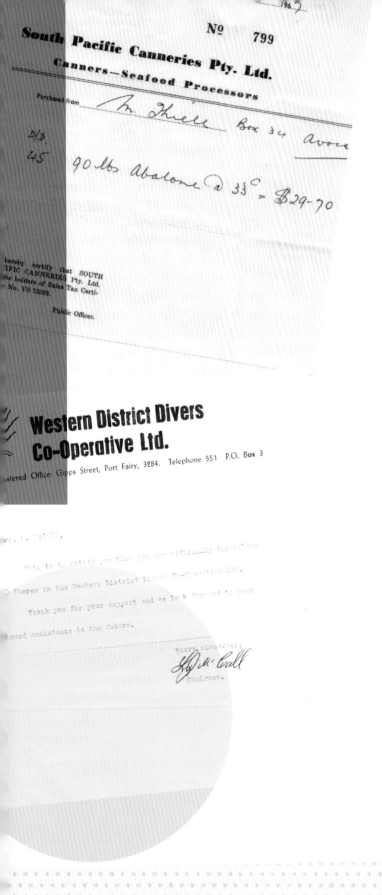

Western District Divers Co-Operative Ltd.

stered Office: Gipps Street, Port Fairy, 3284. Telephone 551 P.O. Box 3

r. I. McC...,

...ote is to notify you that you are officially the holder
... Shares in the Western District Divers Co-Operative Ltd.

Thank you for your support and we look forward to your

...nued assistance in the future.

yours sincerely,

Chairman.

the bends.'

Bob had not only surfaced too quickly, but he had spent too much time at depth. By the time the anchor was up, he was suffering severe stomach cramps, and dizziness set in within half an hour. Despite breathing pure oxygen on board, by the time they arrived home Bob was unable to move his arms or legs and was almost blind. A plane equipped with a portable decompression chamber was sent from Morwell and transported Bob to Royal Adelaide Hospital. He arrived within three hours of the accident. During the week Bob spent in hospital, three hours a day were in a decompression chamber. At the time it was said that Bob had a 50:50 chance of survival without a decompression chamber. The article concluded: 'Against medical advice, Bob Ussher has returned to professional diving.'

Some content derived from *Co-operative Resource Management in Action*, Entity Films 2005.

The Victorian abalone industry

1962	The abalone fishery commenced
1965	Divers peaked at around 300
1965–66	Harvest 376 tonnes live weight
1967–68	Harvest 3384 tonnes live weight
1968	Fishery closed to new entrants. Up to 350 divers. Introduction of two management zones: Eastern and Western. 162 divers LML introduced. Blacklip 10–12 cm, greenlip 13 cm
1969	A $200 abalone fishing licence introduced as part of the Fisheries Act 1968. Introduction of the Central Management Zone. Licence holders: Eastern Zone 34, Central Zone 56 and Western Zone 18
1982	Licence holders reduced to 90 through natural attrition
1983	Further effort made to reduce licence numbers
1983–84	Live weight $4.59 kg
1984	Abalone licences become transferable on a 2:1 basis. Cost $100,000–$160,000 for two consolidated licences, plus $10,000 licence transfer fee. Fees based on average price/kg
1987–88	Live weight $15 kg. Some divers landing 40 tonnes annually
1988	1:1 transfer of Abalone Fishery access licences: 23 in Eastern Zone, 34 in Central Zone and 14 in Western Zone. Introduction of quotas and TACs: Eastern Zone 460, Central Zone 700 and Western Zone 280 tonnes

1992	Introduction of $20,000 abalone processor fee, resulting in processor numbers dropping from 42 to 14 overnight
1995	Increasing illegal activity. Leased licences: Eastern Zone 17%, Central Zone 65%, Western Zone 43%
2000	Abalone thieves activity increasing, having significant impacts
2001	LOFM research confirms that LML not adequate to protect abalone stocks from over-harvesting or serial depletion. Continued competition from abalone thieves. VADA spends $140,000 on a Marine Stewardship Council (MSC) pre-assessment report to identify and highlight ecological concerns. TAC in the Central Zone drops for the first time in 15 years, because of these concerns
2002	Introduction of MOU agreements to spread fishing across the Central Zone. TAC drops in the Central and Western zones for second time in 16 years
2003	TAC drops for the third time in 17 years: total of 191 tonnes having been lost in three years
2004	MOUs start to address effort distribution. TACs stabilising as effort and voluntary size limit increases start to show stability to the resources. Continued problems of significant illegal activity by abalone thieves
2005	New indictable offences aimed at stopping abalone thieves and traffickers. New regulations to stop abalone thieves who claim to be recreational divers
2006	Number of identified areas in the MOU expanded from 12 to 15

3. Western Abalone Divers Association (WADA)

The Western Abalone Divers Association represents the interests of divers and access licence holders across the Zone. Many of its members are those who pioneered development of the fishery and are still active in the industry. WADA supports the divers of the Western Zone and the Association strives to put the interests of the resource first.

The owners of the 14 Western Zone licences have controlled access to the abalone resource, amounting to some 200 to 300 tonnes annually. Each licence allows a diver to collect a set amount of abalone each year under the quota system, which is reviewed annually and varies from year to year.

WADA plays a key role in managing the Zone's abalone resource and protecting the interests of licence owners and their divers. WADA has developed a reputation for putting the interests of the resource first and has been at the forefront of co-operative resource management in Victoria.

Co-operative management has evolved in the Western Zone from a grass roots level. Once there was evidence of the resource being depleted, the owners and divers soon realised that their desire for responsible stewardship could only be fulfilled through a collaborative approach at all levels. This meant open and honest dialogue with all parties, including the scientific community and the state's Fisheries Department. Through goodwill and perseverance, relationships of mutual trust and respect have evolved, which today makes the Western Zone fishery a shining example of co-operative responsible management in action.

Fine-scale management

For management purposes the Western Zone is divided into 36 reef codes. The principles of fine-scale management includes the need to manage abalone product from each individual reef code. This has been a revolutionary concept for abalone production in Victoria.

Traditionally, outputs have been set as a total allowable catch for the Zone. Now, through WADA's initiative, each of the Western Zone's 36 reef codes has quota reference points that are set annually and are constantly monitored through a memorandum of understanding. The abalone harvest from the Western Zone is now evenly distributed across all productive reefs.

The powers behind WADA

Rick Harris

Divers were caretakers of the resource

One of the pioneers of the abalone industry, Rick Harris began his diving career after he sank a small boat off Parkdale, a Melbourne suburb on the edge of Port Phillip Bay. He hired a diver to salvage the boat and became interested in diving for abalone through talking with him. Rick went out and bought the gear, plus the $6 commercial fishing licence.

Rick was a plasterer by trade and continued working at that for two years, diving for abalone on the weekends and working at his plastering job during the week. His dive partner, Noel Middlecoat, was also his plastering partner. Noel lived at Altona and they started diving for abalone on the inshore reefs there. 'I got into it because of the money,' Rick explains. 'Lifestyle was a factor, but it was mainly the money. We could make a week's wages in one day. I'd work all week plastering, and then go out diving for one day and make the same amount of money.'

They went snorkelling for abalone with a square kerosene tin and sold their catch to Les Tuckey in Maribyrnong. They dived at Half Moon Bay near Blackrock, on Port Phillip Bay, and then at the air force base at Altona: 'I called it GI Joe,' he says. They tried right round Port Phillip Bay. The abalone were good at Half Moon Bay, at the wreck of the *Cerberus* and at the back of

the Sandringham Yacht Club. 'In the early days we worked the back of Portsea — the quarantine area. But the army booted us out. We then walked around the rocks on the shore at low tide.'

The water in the bay is very cold, particularly in winter, and they soon discovered their wetsuits weren't good enough for the icy waters. So they made their own. 'The Espies [divers] made their own suits and that was the first time I ever saw a top with a hood and no zips,' Rick says. 'I went to some blokes who made wetsuits and told them I wanted the thickest neoprene they had. A top with a hood attached and no zips. They said it wouldn't work, that I wouldn't be able to put it on. I told them that was my problem. Mine was the prototype. They didn't last very long — I chopped the knees and elbows out pretty quick.'

Rick made lots of experimental suits with a chap named David Dunn. He glued two lots of neoprene together, lining to lining,

so he had two outsides to the suit. But it wasn't a success because the water saturated the lining, which then came away from the neoprene. When they started diving in the deep waters at Portland off the lighthouse at Nelson, they found the neoprene would compress and collapse and stay that way. 'It was useless and so darn cold.'

Then Rick asked his wetsuit maker to glue two lots of neoprene together, rubber to rubber, with the lining worn on the outside and no lining inside. 'I had to use goanna oil to slip into my suit. I also used it in my ears to stop infections. Those suits were really good and you stayed warm. They didn't collapse with deep diving.'

Rick credits Phil Sawyer with inventing the hot-water-bottle wetsuit using two hoses. 'He was lucky he didn't burn himself,' he says. 'We tried everything. Masks with purge valves were popular for a time and big flippers.' They didn't wear boots for a long time and gloves were the gardening variety, not neoprene.

They always wore two weight belts with 36 lb (17 kg) of lead around their waists with the hose tied onto one belt with a cord.

After diving for 18 months in Port Phillip Bay, a few of the divers told him about the abalone at Portland. 'In 1968 I ventured all the way down here in an old FJ Holden with Noel Middlecoat. That's how we got here in the first place.'

Working in the Southern Ocean was new to them after the relatively calmer waters of Port Phillip Bay. There were a lot of big boats at Portland and Rick got a job working for Tony Jones on his 19-foot (7-m) boat with Paul 'Noddy' Hill. He worked on Dick Kelly's 17-foot (5.18-m) Hercules until it sank at Crayfish Bay. Rick explains: 'I got turned over by a big wave and couldn't get out of the break. My deckhand Richard Lakey was in a bit of trouble, so I had to save him. But we lost the boat and everything.'

According to Rick, some of the men had abalone boats but their hearts weren't in it. 'They saw too many shadows, feared sharks too much. Regular hard-working divers seldom saw sharks. Lots of those on the fringe saw them all the time. They were the blokes who always came home with small loads. I don't blame them. You gotta be a water baby if you're gonna work in it.'

Rick recalls how they would all get up in the morning and meet at the quarry carpark overlooking The Passage and decide if they were going to work or to the pub. 'We spent a lot of time in the pub because of the big seas we get down here. We'd also congregate at the boat ramp. There would always be blokes trying to talk us out of going to work. They would try to block the ramp, just sit there with their boats on their trailers and stop the leaders from going by talking and talking — because if one went, then most would go.'

Another ploy to stop others working was when some of them developed bad ears after a good day's diving. Instead of staying home and getting over it, they would go out and sit on their patch and not dive. That way they were protecting their hot spots until they could get back there. 'If I had a good day,' Rick explains, 'I'd never lead others back to it the next day. I'd take them miles away. At the end of the day we would tie up floats on an anchor where the going was good and it made it a lot easier to find the next day. We'd leave them there and pick them up next dive.'

Some of the divers always went to the

same spots and they regarded these patches as their own. Rick always tried to find patches of clear water where he always did well — his theory was that abalone helped to clear the water.

Rick was always a late starter. He had a theory that if he gave the other divers two hours head start, they would get more bent than him, especially if they tried to stay as late as he did. He also reckoned the sea temperature was warmer later in the day. 'I'd often go out at midday and as I was going out some would be coming home. They'd had it. I liked working the late afternoons — I reckoned it was often the quietest part of the day.' As he says, 'Some of the real "hungries" used to work on Christmas Day and Good Friday. They worked any day they wanted because they were their own boss.' But the divers looked after each other, like when Noel Middlecoat broke his leg at Port Fairy after a big wave hit the boat. 'He had no compo [workers' compensation] or anything and all the divers chipped in. Every time we worked, we'd put in a bit of money to keep him going. There were only about 20 to 30 divers in Portland and most of them put in, apart from a few divers who were too bloody greedy. Ron O'Brien ran the account. But most put in what we'd agreed to as part of the Portland Divers Association.'

Rick considers the divers then were caretakers of the resource. They dobbed in poachers and always checked out strange boats on the water. He recalls divers jumping on a boat that was fishing undersized abalone in the shallows of The Passage and dumping his catch overboard. 'It wasn't me but I agreed with what they did because someone's abs were sent back from the Melbourne processors because they were so small they were falling through the mesh in the drain trays.'

At one stage the divers collectively agreed to give The Passage and inside Nelson a rest. Rick suggests this was an example of the divers being responsible resource stewards. But there were always a few who wouldn't go along with it. They tried closures. It would work for a bit but then someone would jump the gun. The divers tried to bring in boat limits similar to Mallacoota but the 'hotshot divers' wouldn't have a bar of it. Rick was pushing for 500 kg as the limit. He liked it when Sou'west Seafoods brought in red light days and quota days because the factory was in effect making sure they got the best return out of the resource.

He recalls having arguments with some of the divers they suspected weren't sticking to the size limit. 'I confronted one of them on the wharf. He said his deckie measured them all. He showed me his gauge. It was an eighth of an inch under. The steel was light gauge — you could bend the prongs in.'

Sometimes there were serious repercussions such as the time Bernie Morton's boat disappeared after he broke an agreement not to work inside Nelson. 'He reckoned he'd been away when the agreement was made so he was not part of the deal. He worked the ground and his boat disappeared shortly afterwards. That was the rumour anyway.'

In the early days a lot of divers got the bends. Rick says they would ride their parachutes up to the surface like rockets from 30 metres. But he figured that wasn't smart so he slowed down, especially after he got bent. Rick's routine was to do 10 minutes' bottom time at 30 metres and then count 100 breaths at 3 metres on the ascent. 'You can sense the bends. You can feel the niggle in your shoulder, the pins and needles in the last 3 metres of the ascent. You'd try to work to the diving tables, but you wouldn't get any abs that way. Everybody got bent. Everybody.'

According to Rick, that's the way diving affects the divers. 'At the end of the day you're snitchy with your deckie. You are sensitive to sounds and noise. I remember I used to be so cold sometimes that I couldn't unlock the car door at the boat ramp. I had to use two hands to hold the key and turn it. I had no strength in my hands and feet and I used to worry about breaking my ankles.'

Some days the local pubs would be crowded with divers. They would get drunk and be raucous and loud — harmless but noisy. Local legend has it that one particular sheller was barred from every pub in town. Rick recalls that a chap named Duncan and 'Ten Bins' (Gary Watson) used to see red when they saw each other. 'They'd have screaming, yelling matches and would want to fight everybody. Whenever we had divers' meetings, Ten Bins would try to run the meetings and would argue. He was a loudmouth, he always had something to say and he'd argue with everyone.'

One of Rick's best memories is coming

home in the dark with over a tonne on the boat. 'We'd come home in storm season in the dark. Coming back in rough seas, we'd have to follow the coast, you couldn't rely on a compass.'

The divers would race each other out to work and then race each other back home. 'You'd see three or four boats racing, they'd be side by side doing 30 miles [48 kph] an hour. You'd think they're going to land on the beach, but they'd just glide in.' In the old days, there was no speed limit and they'd come roaring right up to the harbour, judging it perfectly. Just like a jet landing at an airport. They were experienced drivers who knew the waters and they'd stop on the side of the wharf. 'We'd plane right up and glide into the wharf just like that. You're not allowed to do any of that now.'

David Forbes reckons one of the funniest stories about Rick was at a zone meeting. 'He was having his say about how the young divers don't know anything about anything, and how they all dived too shallow for abs. Then someone asked him how deep he dived, and he replied, "I don't know. I just go to the end of me hose. Not sure how long that is." Turns out Rick pinched his mum's garden hose and would use that.' There's also the story about Rick being seen in Portland driving his four-wheel drive with a cattle trailer and five or six cows on board. The locals say that Rick thought the cows were bored so he was taking them for a drive.

Rick now divides his time between Portland and Coffs Harbour.

Len McCall
Young blokes with no brakes on

It was all about the lifestyle when Len McCall started in the abalone industry in the 1960s. It attracted young men who liked diving and being around the water. Some had crossed over from surfing, as diving gave them something to do when it was too rough to surf. Some lived on the beach and spent the money they earned on booze and girls. As Len explains, 'We were a bunch of young blokes all eager to get out there and test ourselves with literally no brakes on, and we could make good money. I thought I'd do it for a couple of years and then go back to a regular mundane job. Fortunately, that didn't happen.'

Len's lifetime involvement with the water began with swimming when he was very young and living in South Melbourne. 'I

CRAIGS 44
MILLS 263
J.B 206
LEVISE 926
CRAIG 499·5   15TH   R/CRAYFISH
KILLARNEY 205·5   17TH
KILLARNEY 367·5   18TH

ABALONE
KGS. LOBSTER 3.40 PER KG.
365·5
9 BINS
TOTAL $ 1242.70
SIGNED
SIGNED   No. 0297

Clockwise from bottom left: Len
McCall at Port Fairy; page from Len
McCall's books, showing his catch at
various locations and Sou'west receipt;
early air compressor. Len McCall at
work. *Photo source*: Len McCall

wasn't good at sport except for swimming, so naturally I took to the water when I could.' As a child he joined his father pulling mussels off the pylons in Port Phillip Bay.

Len always talked about diving with his mates, but it wasn't until he went to New Zealand in 1965 that he got serious about it. He met up with some guys and one Friday they went out and bought diving equipment. It turned out they were all lying about having previous diving experience. But they went out and tried it, mainly collecting mussels.

Back in Australia, Len did other jobs, but really wanted to get back in the water. His wife Sandra (now Sandra Downes) had a friend who was going to dive at Marlo in the Eastern Zone, and Len went with him. As they drove, Len thought to himself, 'This guy's getting a job as an abalone diver and I'm a better diver than he is.' After a couple of days he went out on the boat and realised he wasn't as good as he thought. 'I was pretty hopeless to be honest. My dive suit was from New Zealand and more suited to warmer waters than cold waters.'

In those days it was easy to get a licence, but Len remembers that most of the Marlo divers didn't have one. 'The Fisheries people came around and said, "We'll be back on Friday, so you'd better have all your paperwork ready and we'll sign you up." We paid about three or five dollars, a very small amount.' This licence meant Len was now fishing commercially for abalone.

The story of the men using sanitary napkins happened at Marlo. In those days the divers made their own compressors, which were called 'hookahs'. They were primitive and required filtration. They used a 3-inch (8-cm) diameter water pipe and put a cap on the top and one on the bottom. It would drain at the bottom and air would come in and circulate around. Water would come out of the suspension in the air, collect on the top and drop to the bottom, and then air would go out through the hose at the top to the divers. It required some kind of filtration system. Someone came up with the idea of using sanitary pads as filters as these were readily available, cheap and disposable. 'They'd get wet and oily, so you could throw them away.' The story goes that the divers went to the general store at Marlo to buy provisions. They bought all sorts of supplies including lots of packets of sanitary pads. They went back the next week and bought a heap more. The guy behind the counter finally asked, 'Just how many women have you got at that camp?' And they replied, 'Only three!'

Some of the divers who weren't married had what Len refers to as 'camp followers' or groupies. The girls were attracted to these young guys who had an intense work life, made big dollars and had lots of free time. 'They were mostly nice young girls out looking for a good time,' says Len. 'The guys were too, and they had the money to do so.'

The market for abalone in Australia developed almost overnight. The entrepreneur Cecil Chang advertised for divers to harvest abalone. Len explains, 'He paid us very little, but none of us cared very much because it was a lifestyle thing. It all started then and other people began to notice as well. There was a demand for it in Asia as well as from the local Asian people.'

As well as Cecil Chang, who helped develop the Asian market, other early buyers/processors were: Smorgan's, South Pacific Cannery, Russell Crayfish, Safcol and Blackney's in Geelong. But the quality of the abalone was not consistent and the handling procedures were primitive. Sometimes the product would sit on the back of a truck for days in the sun.

In 1967 Len decided to buy a boat and dive at Phillip Island. He had dived there previously and knew there was plenty of abalone. He teamed up with Peter Cornelius and they struggled along, not catching much. Len didn't have kids at that time so he could stick it out and bit by bit the fishery gathered momentum. Other divers came to Phillip Island and, although Len dreaded them arriving because he thought they were stealing his fish, he realised it was a good thing that they did because they had contacts elsewhere, plus the volume of the catch increased and people were interested in buying it.

Len describes the early days as being similar to a gold rush. The fledgling industry attracted a lot of different people from different backgrounds. There was a group of divers Len met at Phillip Island who were called 'the longhairs': 'They were the real hippie types who were stoned out of their minds most of the time. They had one of the early aluminium boats which, instead of welding, they had riveted, and all the rivets had popped, so as well as two divers on

board, they had three guys bailing water to keep this thing afloat.' Whenever they had a good day, they would get their money in the quickest time possible and wouldn't work again until they were broke.

While at Phillip Island, Len teamed up with John O'Meara, who was from Penshurst near Port Fairy. 'He kept on telling me how much abalone was down there and how good it was. I was quite happy at Phillip Island and I'd probably still be there today if not for John.' Len and John decided to go to Portland because they knew other Mallacoota and Eden divers there. They stayed overnight in Warrnambool and the next day went out to

the breakwater, where they saw people diving and knew they were catching abalone. They went on to Portland and found there were about 35 boats with two divers on each.

'That was an awful lot of people diving,' Len recalls, 'and the caravan park on the crest of the hill was full of abalone boats.' On the way back to Phillip Island they called in at Port Fairy and talked to some of the local cray fishermen. 'They said there'd been divers there but they hadn't stayed. They'd moved on saying it wasn't very good.' Len and John stayed overnight in Port Fairy and went for an exploratory dive at Killarney, which they thought looked good. Then they walked into The Crags and knew it was good. Just snorkelling off the shore they could see that the relationship between cray weed and abalone was similar to that at Phillip Island. They came for six weeks initially but then moved there for good. That was around 1968.

'We decided to move to Port Fairy and give it a try as John was keen to be near his family. It proved to be an excellent move and we were able to access storage at the Port Fairy Fishermens Co-operative.' They also decided they wouldn't use the mainstream processors like Safcol because, as Len explains, 'We were doing pretty well and we didn't want everyone else to know.' They began selling their product elsewhere.

Len knew the word about Port Fairy had spread when Dick Cullenward, an American he'd met at Phillip Island, came into the pub one Friday night. Len recalls Dick saying, 'Wow Len, I knew you were diving here and I've come down for a look, but I'm not going to ask you what it's like.' Len replied, 'It's not good, Dick,' in an attempt to put him off. But Dick came down to Port Fairy. He was a bit older than Len and had been flying jet planes in the Korean War. 'He was a good diver,' says Len, 'some say he was the best, but it depends on who's telling the story. He was a Yank, he was loud and he was flamboyant, but he was a nice bloke.'

Initially the people of Port Fairy didn't like the abalone divers. There was a certain amount of hostility between the established local fishermen and the abalone divers, who came along with their basic equipment and made heaps of money. After a while the divers employed some of the locals as deckhands and they'd shake their heads at what the divers were doing. But gradually the divers won the locals over and became part of the community.

When licences became limited, this caused more friction with the locals. 'Only a certain number of licences were permitted,' Len explains, 'and once you locked into that, you were there and those on the outside looking in were jealous and envious saying, "Why you and not us?" There was one guy in this town who complained, saying he'd been raised there in a fishing family.' According to Len, he did a bit of diving but he didn't get off his backside and get a licence.

The licence fee went from $5 to $200, and then to around $57,000, which was the licence fee plus a royalty. Len explains that when a royalty was first suggested it wasn't a popular thing, but they gradually got used to the idea as it would underpin the industry. The royalty was based on the beach price of the previous year and came out at about 7.5 per cent. 'It was the royalty we paid for access,' Len explains. 'Because we don't own the resource, we have an access permit.'

Sou'west Seafoods began in Len's rumpus room on a very cold, wet day. Len explains, 'We had a week of bad weather and John O'Meara had come around to talk about marketing.' They had been storing

the product in the old Fishermans Co-op building and Len would phone buyers and let them know what quantity they had available and get some leverage. Gradually the process developed and John and Len decided to form a co-operative of divers. 'That started the ball rolling. I jumped on the phone and talked to all the people in the group. We needed a certain number, so we invited Red Quarrell, who was in Port Campbell.'

But because they still needed more people, they incorporated their wives and girlfriends as well. 'We had Murray Thiele and his wife Esme, Andrew Coffey and Valmai, Dick Cullenward and Natalie, John O'Meara and Maureen, Red Quarrell and his girlfriend who became his wife, myself and Sandra. We got together and called ourselves

the Western District Divers Co-operative.' Len did the marketing and was the chairman. They bought out the old Fishermans Co-operative building on the wharf at Port Fairy and changed their name to Sou'west Seafoods Co-operative Limited.

Initially the abalone was shelled at sea but when the regulations changed they had to bring in the whole catch in the shell and shuck it in the Co-op shed.

As time went on they decided to drop the co-operative model because, Len explains, 'Co-operatives, especially in the fishing business, have a high rate of failure mainly because the members drain too much of the resources out of the company and leave it with nothing to run on. We adopted a corporate model and called it Sou'west

Seafoods Pty Ltd.'

Len dived for 31 years and says the most startling thing he saw was when he was diving at 12 metres off Killarney. 'There were schools of fish swimming in the crystal clear water when suddenly there were jet-like bursts streaming into the water. It was gannets and they were dive-bombing for fish. I could see them going underwater down to 9 metres. I've never seen it again.'

Len says the abalone industry has been good to him. 'I'm not that well educated and it's given me an opportunity to learn an awful lot of things, particularly in the corporate world. So if anything I owe the industry more than the other way round. It wasn't all that easy in the beginning. There are people who say that it was a Cinderella industry and we all just fell into it and the money piled in. There have been peaks and troughs and we certainly kicked a few goals in our time. But bringing all these individualistic and highly competitive divers together to work as a team was always a vision of mine. Without that happening we would likely have remained just a group of individual Western Zone abalone divers.'

In October 2011 at the Victorian Seafood Industry Awards, hosted by Seafood Industry Victoria (SIV), Len was awarded the prestigious Seafood Industry Icon Award for his 45-year involvement in the industry as a representative at all levels. The industry acknowledged that, as a diver in the Western Zone, Len was a driving force in establishing the Western District Divers Co-operative. And as chairman of the Victorian Fishing Industry Federation, he led the transition of that organisation into what is now Seafood Industry Victoria.

In 2013 Len was inducted into the National Seafood Industry Hall of Fame for his significant contribution to the National Seafood Industry.

Lou Plummer
An abalone dynasty

Lou Plummer is passionate about the abalone industry. Abalone is a family affair in the Plummer household and Lou's very proud that all his children are involved in his business. Sons Phillip and Glenn are each nominated divers and dive on his licences. Daughter Joanne is a licence part-owner and also does the computer-based bookwork while her sister Terry handles the mail. Lou is also quick to acknowledge the support given by his wife Shirley, who was always there to support him and raise their family: 'She's the love of my life.'

Lou was working as a painter in the mid 1980s for $400 a week. 'I hated being a painter. I was employing three painters, but people wouldn't pay me and the fumes made my chest bad.' Lou used to go spearfishing with Murray Thiele. At the time he thought he'd like to be an abalone diver but there were no licences available. They did some abalone diving in New South Wales but the Fisheries people there wanted to get rid of casual divers. '"We bombed out there because we weren't locals,' he recalls.

On weekends Lou was also making money from collecting periwinkles (another marine gastropod) at Port Fairy. He would collect up to 300 kg and on Sunday nights he'd pack them into cages and drive to Geelong, where he'd hang them over a pier to keep them alive. On Wednesday nights he'd drag them out of the water, take them home and bundle them into 20-kg bags. At 5 am on Thursday mornings he'd drive to Melbourne and sell around three-quarters of his catch to the Victoria Market, then the rest at the Camberwell and St Albans markets. 'I'd

be back in Geelong by 10 am with maybe a couple of hundred dollars spending money in my pocket. I'd then work an extra two hours that day to make up for time lost.' This was good money in the 1980s and Lou continued to do this for a few years.

But the lure of the abalone industry was always there. 'Murray made reasonable money from abalone diving. I thought I could work as a painter during the week and dive for abalone on weekends.' Within a fortnight Lou knew he wanted to become a diver and earn $400 a day, instead of a week.

A Western Zone abalone licence became

available in 1985. Lou wondered why the other divers weren't buying up licences to keep newcomers like him out, but he jumped at the chance to get out of painting. In the mid 1980s you had to buy two licences then submit these to Fisheries, which would cancel them and issue one in return. The bank loaned Lou $200,000 to buy two licences and a 17-foot (5.18-m) aluminium boat, a compressor and an air hose, plus an old Toyota tray body that had been sunk a couple of times and was covered in rust.

'I didn't know where to start. It was hard work. The first day I set up the compressor on the beach because of the rough weather. It was as rough as billyo and waves thrashed around me. I kept swimming around in circles and thought someone else had been on that patch — but it was me. I just didn't have the knack of going this way and that. When I knocked off for the day with a catch of 57 kg I found I was the top diver in the area on the day because no one else had

worked because of the rough weather. Then it occurred to me that I'd just borrowed $200,000 from the bank and had to pay that money back, but no one was working. I was used to going to work from Monday to Friday, even if it was raining.'

It took about six months for Lou to learn what he needed to know about abalone diving to compete with the experienced divers — the 'guns'. He liked working the bays at Port Fairy and while the other divers used big boats he decided to adopt a different approach. The guns were working one or two days a week but he thought he could do even better if he went out more often in a small boat. Friends owned land with access to local bays and he began launching his boat from their properties. While the guns would work about 40 days a year out of their big boats, he'd work around 100 days from his little boat as well as 40 days from his big boat.

Lou had a bank loan for five or seven years and he worked hard, paying it off in 15 months. While Shirley kept the family going in Geelong, Lou lived in a caravan at Port Fairy when he was diving. It wasn't too long before Lou's younger son Phillip decided to join his dad in the business but, 'Phillip wanted to have girlfriends around, so he lived in a tent.'

In 1992 Lou's elder son Glenn wanted to work in the business too, but they really needed another licence to support the three of them. Fortunately Lou happened to meet up with an older diver, Dick Kelly, who wanted to sell part of his licence. 'Two or three days later Dick came back to me and said he wanted to sell *all* the licence,' Lou recalls. 'I paid $1.3 million for it when 18 months earlier it would have cost $180,000.

While this was a huge capital investment, Lou understood the value of having another licence because the beach price was going up every day. 'When I started, the beach price was $2.40 per kg. Then every day I dived it went up, say, another 20 cents and kept on going up and up and up. Andrew Coffey and I were really worried it would hit $10 per kg. By the time I bought the second licence it was around $35.'

Lou claims that in the early days he could last without a wetsuit for 20 minutes. 'When I first started I bought a thin wetsuit and it was so thin I'd be shaking with the cold. Andrew Coffey told me he used to warm himself by wearing newspaper under his wetsuit and so

I did this. It worked for an hour or two, but God it was a mess when I took off my wetsuit and I was covered all over with newsprint — it didn't come off for weeks.'

There were no quotas in those early days, so pickings were rich. Lou recalls he only had a small boat at the time and would go diving with Phillip. 'I'd jump out of the left-hand side of the boat and go like billyo for about four hours. Then Phillip would get into the gear and go like blazes out of the other side of the boat for a few hours.'

Lou was spending anything up to 140 days a year in the water and it started to take a toll on his health. He began getting sick. Divers didn't know then about the benefits of taking oxygen when they finished diving for the day. Standard practice was that if they dived in deep water, they'd move to shallow water for a time, then surface and go home. 'Then we found that if we took a swig of oxygen for 10 to 15 minutes on the way home, we'd be as good as gold.' It soon became common practice for divers to have an oxygen cylinder at the front of their boat.

'I'd been diving maybe five years and was probably pulling in 40 to 45 tonnes a year on my own. That's a lot of fish. When quotas came in I had to back off and this meant I had to have a lot of days off. But what happened was that if I had a few days off, I'd be really sick for the first couple of days I went back into the water. I'd be at the bottom working and getting rolled about and then I'd feel no good. I'd make a dash for the surface so I wouldn't be sick in my regulator, because I knew I'd have to take the sick in again in my next breath. I'd manage to get to the top and then I'd "let it go" before giving the mouthpiece a wash and off I'd go again. An hour later I'd be up again doing the same thing.'

One day Lou was diving at the Port Fairy lighthouse and became violently ill and raced to the surface. Out popped his bottom dentures and he went straight back into the water to retrieve them, but had no luck. 'I jumped back into the boat and headed to where Len McCall was diving close by. I asked him if he'd seen any "biteys" and he thought I meant sharks.'

Lou lost about 2 kg a fortnight because of this sickness, which was later diagnosed as vertigo and was a major problem for him. But, as he says, 'I soon got sick of being sick.' It was 1995 and, with his sons eager to work

in the business, Lou realised it was time to rationalise their operations.

The first plan was for elder son Glenn to be Phillip's deckhand but, according to Lou, 'The little buggers used to argue. Two brothers — oh boy. They were at it all the time. It just didn't work with one strutting around being a diver while the other was the deckie.'

Lou's next-door neighbour Herbert Waldron found himself out of work around this time and Lou took him on as his deckhand. With the tension between the boys, they swapped and Herbert became Phillip's deckhand and Garry Dempsey worked as deckhand for Glenn.

This meant each son was working a licence and Lou was out of the water. But Lou laid down the rules. 'I wanted them to do their own thing. One worked one licence and the other worked the other. I encouraged them to go out at different times and not compete with each other as I knew they'd be at loggerheads.'

Then in 1994 diver Peter Ronald phoned Lou and asked if he had a diver's job going. As it turned out, Peter had an option for a licence but wasn't going to buy it. By then licences were worth around $5 million. Lou rang Murray Thiele, whose response was that he'd 'always wanted to do something out of the ordinary'. Knowing it was difficult to service a licence worth that much, they agreed to go into partnership and buy the licence. Two days later Lou rang Peter Ronald and said they would buy the licence and wanted him to be their diver. Lou and Murray's instructions to Peter were, 'Here's the licence. Here's the quota. Do what you've got to do and we'll keep out of it.'

Abalone divers are an ingenious lot, with most making their equipment out of whatever they could find. 'Most made their own compressors because the gear just wasn't around and it was pretty pricey. While the engineers made up proper stainless steel fittings and tanks, we'd go down to the milk bar and wait till they'd throw out the large soft drink containers. We'd drill a hole in these and that was our reserve tank. They still use these cans for this today.'

There was a lighter side to diving for Lou. He tells how Lady Julia Percy Island is notorious for sharks, especially white pointers. 'When we went to the Island some of us would wear football jumpers to ward off the sharks. I wore a North Melbourne one with blue and white stripes. I don't know whether it did any good, but we were told it made us not look like "bitey" food. Tourists used to come over to the Island for fishing and would want to know why we were wearing footy jumpers. We told them we got cold and when we got to the Island we'd kick the footy.'

Lou tells the story of the most expensive driveway he ever bought. It was 130 acres (53 hectares) that had access to the bays by land and by a small boat and he bought it at auction in the 1990s. Nearly 20 years later Lou and ten others turned this land into the site for Southern Ocean Mariculture Abalone Farm.

Lou still loves the industry, but life is much more relaxed for him and Shirley these days. They spend a lot of their time travelling, enjoying life and enjoying each other's company.

Murray Thiele

Diving was like a drug to me

Murray Thiele first started fishing for abalone in 1964 on a fishing licence that cost £2. A true pioneer, he was in the first group of divers to purchase a licence specifically for

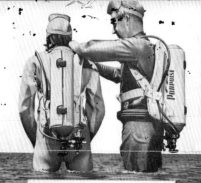

Clockwise from top left: Murray Thiele; hoisting a catch; early skin diving brochure. *Source*: Murray Thiele

abalone in 1969 and has held one or more licences ever since. Murray first became aware of diving during World War II when he saw photographs of Defence Force frogmen.

Three things happened to Murray in 1958: he purchased a sawmill in Avoca, his wife had a baby and he bought an aqualung from the Melbourne Sports Depot for £67 18s 6d. This was a huge financial outlay in those days when the basic weekly wage was £12 16s, so it was more than five weeks' wages.

It took Murray two years to master the breathing technique. He'd always been told you couldn't breathe underwater but it just didn't seem right to him. He dived with aqua diving clubs and formed one in Avoca. 'In 1963 our club decided it needed some money and that they would get some muttonfish [abalone]. About six of us went to Port Fairy but the sea was as rough as bags. No one went into the water and we didn't get any ab.'

In Easter that year Murray went to Port Fairy again and dived for abalone for three days. Pleased with his catch, two days later he sent his catch to Melbourne by carrier. 'Some had gone off and were bad, but I still earned £53. So I thought, bugger it, this isn't too bad

at all.' He continued to run his sawmill and dive for abalone at Port Fairy on weekends. 'When I was in Avoca I was a sawmiller and when I was in Port Fairy I was an abalone diver. Sometimes I'd get to Port Fairy only to get a message that there was a problem at the mill and I'd have to go back home again without diving.'

Murray's part-time diving didn't sit well with the other Port Fairy abalone divers, who labelled him a 'weekender'. Murray recalls there were a lot of cowboys in the industry in the early days in Port Fairy with around 500 people getting abalone — some with licences and some without. Many came from Mallacoota and they earned good money. But a lot of them lived it up. When Fisheries decided to create an abalone industry, Murray thinks the reason is that the divers were seen as a nuisance. Licences went from costing £2 to $6 a year, and then in 1969 the annual licence fee skyrocketed to $200 .

Of the 500 or so divers, only about half took up a licence once it became $6. Then only 120 remained when the price went to $200. According to Murray, 'You didn't sell a licence in those days. If you wanted to get out of the industry, you just left. The licence fees varied over the years depending on the beach price. One year I remember paying $68,000 for a licence for 12 months.'

But the Asian market for abalone was escalating. In 1965 Len McCall rang Murray and said the divers were going to form a co-operative and wanted him to join. According to Murray, Len didn't beat around the bush. He said to Murray, 'I don't like the way you work, but you've got a licence in the Zone. Do you want to join our co-op?' Murray recalls they received a 4 pence a lb (0.45 kg) raise because they pooled their catch and sold in bulk, which made it more efficient for the buyers.

There were tensions at times between the men. Murray tells this story: 'Carl Armstrong had a little shed and freezer right on the wharf. Len McCall, Johnny O'Meara, Harry Bishop and Dennis Carmody were the clique and only they were allowed to use it. I was out of it because I was a weekender, even though I was here a long time before them. I couldn't put my fish in their freezer because I was told the others had filled it. Dennis Carmody and Carl had a blue over the price and he said, "All right, I will stick to your prices and I can let anybody use the freezer I like." After that

Carl said I could put my stuff in the freezer — and I did.

'Then Keith Fox, a truck driver, came up and said to me, "I take all the other fish to Melbourne, can I take yours?" I said, "Yes, by all means." I was working with another diver at the time and he was doing the decking. He had a two quid [£2] licence and he wouldn't pay the $200 when it came up. We had our bit of fish stuck in the freezer with all the other divers and Keith said he was taking a load in the morning. We went out diving the next day and all the fish was taken from the freezer except ours. I thought he must have had a truckload. This happened another three times. The others had virtually held a gun to his head and said, "If you take Thiele's fish, then you're not taking ours." So I had to take it myself.'

In 1980 the Fishbrooks factory on the wharf at Port Fairy came on to the market. Four divers — Dick Cullenward, Len McCall, Andrew Coffey and Murray — pooled their resources and bought the lease on the factory. Murray recalls that Len was the driver of this venture, noting that Len was good for the local industry. 'They thought of me as a bushie because I wasn't a gun diver.'

Dick Cullenward pulled out of the co-op

after a while. Andrew's wife Valmai became the secretary and they used to meet once a month. 'We had a few bob and spent money on a development in Geelong. We also bought a block in South Australia and 5 or 6 acres [2 to 2.5 hectares] in Awabi Court, Port Fairy.'

By 1980 the co-operative had become a company and the factory, which was leased, became too small for their operations. They initially wanted to extend the factory on the wharf, which they held on a 21-year lease with 14 years remaining. 'Len was the instigator and got the survey done and did all the work. We looked like it was going ahead, but then Lou Plummer came up with an idea: "Why don't we build on the block of land?" So they did.

'When we first went to Awabi Court there was only one building. Len, my wife Esme and I walked in the front door and Len said, "Gee, why have we bought such a big shed?" But it's been extended three times and it's still too bloody small! In hindsight that move was best thing we ever did.'

Diving for abalone remains a labour-intensive industry. According to Murray, 'In those days I would take the boat out, anchor it, get into the water and get a couple of bags under the boat. Then I'd climb back on board and shell them. Then I'd do it all over again.' Murray has dreamed up all sorts of ways to collect the fish, including using a vacuumer, but as he says, 'You could have the biggest boat in the world and the flattest seas, but you've still got to get into the water and chip them off one at a time and put them into a bag.'

Like the other divers, Murray has built much of his own equipment. 'I'm inventive, I built one of my first boats. It was 13 feet [4 m] long, 4 feet [1.2 m] wide and 10 inches [25 cm] deep. It had handles on the side that were off a coffin so I could carry the bloody thing. It was made of plywood with fibreglass on part of it. I could fit more ab on that boat than I could catch in a day — three bins across and one or two bags sitting on top.' Each bag held about 80 kg and the bins held 45 kg each, around 300 kg. 'That's a lot of abs, especially on that little thing.'

Murray met Lou Plummer at a spearfishing competition in 1970. Lou was interested in diving with the club. When Murray went home and told Esme that there was a new club member, she asked his name:

'It sounds funny now, but I said to Esme I'm not sure whether he's Lou Plummer the painter, or Lou Painter the plumber.'

Murray recalls that 'Lou tried to get an ab licence for years but they froze licences. In 1969 or 1970 we went to New South Wales, where you could get a licence for next to nothing. The philosophy then was that the oceans belonged to the people and if you could make a living out of it, then well and good. If you didn't, then you dropped out. That was a big mistake, because the ab industry got decimated because everyone wanted to get into it. By the time they introduced laws, it was too late. They ruined their own industry.'

While the other full-time divers had deckhands, as a weekend abalone diver Murray could get away with not having one —

and he didn't want one. 'Some had their wives as deckies because of the economics and because she would be available all the time.'

Murray decided to become a professional abalone diver after his sawmill burned down in 1985. 'I was the oldest diver at the time and I dived until I was over 70, which everyone said was too old. It's supposed to be a young man's game, but I've never had any problems. I'm 86 now and still diving a bit with Lou. Why not? Gee, I nearly went out yesterday because I didn't have anything to do, but it was a bit too rough.'

Murray is quick to admit that 'I wasn't the fastest. I wasn't a gun or even a popgun. But while some divers would go out and get 500 kg, I'd go out and get 250 to 300. But I was happy and proved to them it wasn't only a young man's game.' While the other divers used to get their quota in 30 to 40 days, Murray could take up to 70 days to get his. In those years if he didn't fill his quota, he always knew he could lease it to someone else.

Murray's son Cleve took over Murray's licence for seven or eight years and his other son Wayde dived for two years on the licence Murray shares with Lou. Murray is quick to admit that Cleve was a better diver than he is. Murray has been deckhand to both Cleve and Wayde over the years: 'I always said I was the oldest diver and oldest deckie in those days.'

Murray remembers the day Cleve and Wayde were diving and he was their deckie. 'Cleve got into the water, went a few metres, then turned around and got out saying, "There's a bloody great thing just sitting there looking at me." I'd been diving for years and never seen a "bitey" and he sees one after two months. Maybe I just didn't look. Wayde decided he'd dive and went down 10 metres before coming up

with a mask full of blood after a nose bleed, although he didn't see the shark.' Neither son has ventured back into the water again.

The bends? Well, Murray has never had them and believes that's because he didn't dive deeper than 40 feet (12 m), where he says you can stay all day without getting the bends. 'I've been deeper to 145 feet [44 m] but that's not fishing — just down and up straight away. But at one stage, in 1968 or 1969, I was at The Crags [12 km off Port Fairy] and I was diving by myself and had a panic attack.' Not a bad track record for someone who has been diving for nearly 50 years.

Murray jokes that abalone is said to be an aphrodisiac, but reckons 'speaking from experience, that's not true'.

He only eats abalone 'his way'. He gets three abalone, three potatoes, a big onion, garlic and some bacon then puts it all through a mincer. Esme makes up a batter and then makes it into patties. 'They don't taste like abalone but they sure taste nice,' Murray quips.

4. The researchers

The Western Zone is at the forefront of co-operative resource management in Victoria. In this relatively new industry, marine biologists have only recently discovered the true nature of abalone behaviour and reproduction. WADA works closely with the researchers, realising that respectful stewardship of the resource can only be achieved through a team approach aimed at involving both the scientific community, the Victoria Fisheries department and the divers.

One of the first examples of collaborative research was in October 2004 when a major exploratory fishing exercise was conducted within the Julia Bank reef code near Portland. Under the direction of the Victorian Fisheries research arm, Primary Industries Research Victoria (PIRVic), all available access licence holders and divers gathered in Portland along with researchers and Fisheries personnel. Using a predetermined grid system, divers were allocated search areas to explore. Data from the project enabled the catches to be recorded through GPS. At the time the exercise was the largest co-operative research project of its type undertaken in Australia.

The Western Zone is fortunate to have two renowned marine biologists leading research into the fishery: Dr Jeremy Prince and Dr Harry Gorfine.

Dr Jeremy Prince

Jeremy Prince is an internationally recognised abalone expert who consults on the assessment and management of many fisheries around the world. He has worked on

the interface between government and the fishing industry in most of Australia's more contentious fisheries.

Jeremy knows what it's like to be an abalone diver as he has held a licence in New Zealand. He has the utmost respect for the Western Zone divers and believes that they must have immense self-control to go to work every day. 'There are not too many professions where you step into a war zone when you go to work. You spend the day having to control your level of fear because your imagination works overtime — all the time.'

Jeremy began his involvement with the Western Zone divers in 2001 and has been working with them to develop detailed maps of their area and systems for managing the resource at a finer scale. Using their own knowledge of the reefs and their ability to organise, WADA has managed to halt the depletion of reefs that was sequentially occurring prior to 2001 and rebuild stocks following a viral epidemic in 2007.

According to Jeremy, 'Other zones and states are really worried about where their own assessment and management processes are going and are interested in the innovative models being developed by WADA. The second-generation divers from quota-owner families are knowledgeable and concerned about the future of their zone. WADA is seen as the frontrunner of what comes next with abalone. 'An absolutely ground breaking zone.'

In 2001 when Jeremy became involved in the Western Zone, there was concern about size limits. He was asked to do a rapid appraisal of the size limit, which was 12 cm at the time. Complicating the issue was that different trends were occurring on the reefs at either end of the Zone. At the Portland end, which is close to one of Australia's major upwelling areas, the water is enriched and the abalone grow faster and mature at larger sizes. The 12-cm size limit provided little if any protection at that end and by the late 1990s the stocks were in decline. In contrast, the size limit gave greater protection to the stocks around Port Fairy and Warrnambool because the abalone at that end mature at smaller sizes.

Jeremy sat down with the divers in each area and got them to explain and draw their grounds so that he could learn about each of the reefs. There was discussion about introducing voluntary size limits to protect more breeding and to stabilise the stocks where they were declining. The idea was resisted strongly at first, mainly because of a perceived lack of trust among the divers.

After 12 months, it was agreed that the idea should be trialled with a 12.5-cm size limit.

Within nine months the divers noticed greater daily weights, although the number of abalone they caught remained the same. Within the year the divers were asking for larger voluntary size limits to be introduced throughout the Zone. Against Jeremy's advice, size limits were raised. The result was that divers found it difficult to fill the total allowable catch (TAC). WADA learnt from this experience and relaxed some of the voluntary size limits at Port Fairy. They also decided to review and control the level of catch expected from each reef before making further adjustments to their size limits.

Reef assessment workshops

At the end of 2003 WADA asked Jeremy to run training workshops and develop a methodology to facilitate the setting of size limits and catches for each of their 32 reefs. During this workshop participants learned to use trends in catch and the appearance of the shells to place each reef into one of eight categories, according to their state of depletion.

Voluntary limits

The reef assessment workshops became the mechanism that allowed the divers to develop shared views and agree on collaborative strategies. The first workshop in October 2003 set voluntary catch caps and size limits for each of the reef codes, and participants agreed to try and improve access to several areas. They also agreed that WADA ask for a 7.5 per cent reduction in the Zone's TAC: Victoria Fisheries subsequently accepted WADA's recommendation.

When talks first began on the possibility of voluntary limits, Jeremy recalls how each person had their views on the situation. It was a dynamic with which he had previous experience.

The Crags was critical for WADA because most divers used it and that was the area they could come to some agreement about. The first experiment at The Crags made Jeremy aware of something he had not anticipated. 'There is a two to three year period in the life of abalone, around the time when they first mature and come out from caves and crevices in the reef, when they double their weight every year. While the measures taken were aimed at ensuring the long-term future of the fishery by stabilising and rebuilding the amount of breeding occurring, I hadn't realised that a small increase in the size of the abalone being caught had a much bigger impact on their average weight. The result very quickly was a much bigger abalone on average. Within a couple of months the divers could see they were getting better weights for their diving days, by catching the same number of abs.'

Characters in the Zone

Jeremy recalls a minister's adviser saying to him at a government meeting, 'Ahhh, ab divers. When I was a kid they were the ones who would go out and buy a brand-new Monaro one week, stack it over the weekend, then go to town the following week and buy another one.'

The characters of the Western Zone have left a lasting impact on Jeremy and he has many fond memories of the men who he says 'dive long and work hard'. But he observes how the industry has changed from the days of radicals and rascals, the men who had a 'good and short' attitude to life. These days, he says, the industry is more professional, but the buccaneering spirit of the people still exists — and, in fact, flourishes.

According to Jeremy, every fishery has a numbers man, and in the Western Zone the number cruncher is Glenn Plummer. Then there's Lou Plummer and Murray Thiele, who Jeremy affectionately recalls still pull on a wetsuit and have a look around. 'They do it at really pivotal times when the situation is becoming polarised. They will often intervene in a very practical way and they have lots of credibility when they support what new generation divers are saying.

According to Jeremy, Peter Riddle is a top diver. He tells how Peter used to have a fabled greenlip reef where he'd pull a quick 400 or 500 kg a day. It was common talk among the other divers, but no one knew where it was. One day Jeremy went out with Peter to look at the state of the fishery. 'He said that if we didn't tell anyone, he would take us to the greenlip reef — and he did.' Peter found the reef while spearfishing one day and had 'kept it off the radar' from the other divers.

Jeremy also recalls how one ex-diver used to stand outside meetings and smoke joints. 'He'd come in and sound off, and then go back outside and have another smoke before walking back in again.'

Jeremy believes that the abalone industry is a way of life. A fraternity. He says there's something about the creature that gets you in. 'I really miss diving. I don't do crosswords, but the pleasure of jumping in at a new spot and thinking, "I wonder where they are", and then going off to find them. I think it must be akin to doing crosswords. It's just something I enjoy doing.'

Fisheries Act 1995

FURTHER QUOTA ORDER FOR THE ABALONE FISHERY

I, Ross McGowan, Executive Director Regulation and Compliance (Fisheries), as delegate of the Minister for Agriculture and Food Security and having undertaken consultation in accordance with Section 3A of the *Fisheries Act 1995* (the act), make the following Further Quota Order under section 66D of the Act:

[signature]

Ross McGowan

Executive Director Regulation and Compliance (Fisheries)

Date: **26/3/2014**

1. This Order applies to the period commencing on 1 April 2014 and ending on 31 March 2015 ('the quota period').
2. The total allowable catch for blacklip abalone in the western abalone zone for the quota period is 56.10 tonnes of unshucked blacklip abalone.
3. The total allowable catch for blacklip abalone in the central abalone zone for the quota period is 307.70 tonnes of unshucked blacklip abalone.
4. The total allowable catch for blacklip abalone in the eastern abalone zone for the quota period is 417.50 tonnes of unshucked blacklip abalone.
5. The total allowable catch for greenlip abalone in the western abalone zone for the quota period is 0 tonnes of unshucked greenlip abalone.
6. The total allowable catch for greenlip abalone in the central abalone zone for the quota period is 3.4 tonnes of unshucked greenlip abalone.
7. The quantity of fish comprising an individual blacklip abalone quota unit in the western abalone zone for the quota period is 200.36 kilograms of unshucked blacklip abalone.
8. The quantity of fish comprising an individual blacklip abalone quota unit in the central abalone zone for the quota period is 452.50 kilograms of unshucked blacklip abalone.
9. The quantity of fish comprising an individual blacklip abalone quota unit in the eastern abalone zone for the quota period is 907.61 kilograms of unshucked blacklip abalone.
10. The quantity of fish comprising an individual greenlip abalone quota unit in the western abalone zone for the quota period is 0.00 kilograms of unshucked greenlip abalone.
11. The quantity of fish comprising an individual greenlip abalone quota unit in the central abalone zone for the quota period is 100.00 kilograms of unshucked greenlip abalone.

Notes

DEPI is committed to rebuilding abalone stocks in the central zone and establishes the central zone blacklip TACC subject to the following conditions:

1. Industry must manage catch within the target range (between upper and lower limits) for each Spatial Management Unit (SMU). This must include effective voluntary closures of SMUs when a limit is reached. The Department of Environment and Primary Industries (DEPI) will support voluntary management by informing industry weekly of the current catch levels against targets. DEPI will provide daily updates (except weekends) when catch in an SMU is within 10 per cent of its specified limit.

Dr Harry Gorfine

Harry started working on abalone in the early 1980s, doing experimental studies on growth and mortality in Port Phillip Bay. Then, while investigating the diets of abalone, he became involved in abalone aquaculture when this was in its infancy around 1989. At that time there were no abalone farms in Victoria, apart from an experimental farm at Point Henry on the outskirts of Geelong.

'The project I worked on then was funded by Sou'west, so even back then they were looking at aquaculture. Southern Ocean Mariculture has become part of that. I got to know the Western Zone people since they were paying the bills and we were reporting to them on our results,' he says.

In those days, quotas were 20 tonnes per licence and expressed as units: 1420 units, 1440 tonnes in the Victorian fishery. The catches in the Western Zone had been lowered to around 16 tonnes and were up to around 30 tonnes at the other end of the state. This meant some divers were taking large catches while others were taking small. 'Divers lobbied the then premier Joan Kirner. It was a complex issue and the premier made a unilateral decision to set the catch limit at 20 tonnes, which was 70 per cent of their average catch for the preceding five years.'

Harry describes abalone as being 'plastic', meaning they can adapt to their environment when it changes. 'Abalone are fairly good at hedging their bets on the changing environment. They don't all spawn at once, but they do spawn within a period. They don't all grow at the same rate and this isn't an adaptation of the fishery. This is a levelling out of the number being recruited into the fishery.'

Abalone are opportunistic feeders that tend to clamp onto a rock and wait for food to drift by, but they may move daily, seasonally or when food becomes scarce. The colour banding on many abalone shells is due to changes in the types of algae eaten. Juvenile abalone graze on algae and as they grow they rely on drift algae. High wave energy smashes macro algae growing on the reef and the abalone wait in gutters for food fragments to be pushed up the surge channels. Harry says he's seen one abalone clamping onto a piece of food while several others were eating the same piece. Smaller abalone can graze microscopic algae on the surface of rocks.

Abalone is a habitat responder and there are natural phenomenon that impact on the fishery. For example, each year between November and May, the Bonney Upwelling brings nutrient-rich water to the surface and then explodes into a bloom of photosynthesis. This feeds the swarms of krill (shrimp-like food of blue whales) and creates an ocean banquet for fish, seals, dolphins and others. If the upwelling fails, the blue whales don't come in because they can't feed on krill, and the krill can't feed off the micro-phytoplankton bloom caused by the upwelling.

Reproduction and mortality

Harry explains how a female abalone can produce between 20,000 and two million eggs, and how these can either be all expelled at once or in small parcels. 'We know in that early period that it can be highly variable with a high mortality rate.' When the abalone larvae are formed, they have their own yolk sac and it is a high-energy function for the female to output the eggs. Female abalone can respond to cues and reabsorb the eggs to give themselves nutrition, or expel the eggs so that they will settle and hopefully survive the first few days when mortality may be as much as 90 per cent.

Harry believes the Western Zone has the fastest growing abalone and there's plenty of high-value abalone in their natural state, and therefore there's an incentive to fish them. 'All molluscs potentially produce a huge number of eggs and if they all got fertilised we would have massive numbers. But they don't all get fertilised and a large number die. We can still have very high densities, but the densities vary because the environmental conditions change from year to year.'

According to Harry, adult abalone don't have high mortality — rather, high mortality occurs when the molluscs are small and juveniles. As adults, they are strong with a strong shell, which few predators can bore through. If they clamp onto a piece of rock bed they are well protected from most predators, with the exception of Port Jackson sharks and stingrays, which can crush their shell if it is weak. Although worms can bore into the shell and weaken it, this can also cause the abalone to put a lot of energy into the shell to strengthen it.

Working with the divers

Despite all the regulations, Harry believes the abalone industry has been pro-active when facing issues affecting the fishery. 'This doesn't mean they are a bunch of golden-haired boys by any stretch of the imagination. They're a real mixed bag. What it means is that as a group they are different to other fishermen, and this is probably because they are underwater and can see changes to the environment. We've got a fishery that's valuable and quotas promote the idea of stewardship.'

While Harry works with the Zone's divers and stakeholders on a professional level, he also gets to interact with them on a personal level. 'If you walked in on a group of abalone divers, he says, you simply wouldn't be able to work out what their occupation is, unlike with a bunch of surfers, tradesmen or businessmen. You look at the faces and body shapes and there are all shapes and sizes and heights — some fat, some skinny and from all sorts of backgrounds. They're a cross–section of life. Perhaps there are some common themes in attitude and lifestyle, but you can't stereotype abalone divers.'

He recalls how 'One guy's wife used to go off crook because he would knock off the garden hoses to take out diving. This guy's a real character and the story goes that he couldn't go out any deeper than the length of the hose. Another story that does the rounds is that this diver had such a good sound system on his tractor that he would drive it into Portland.' Then there's the story of another diver who used to have a 'dry suit' for diving, but he cut off the legs.

Harry also recalls the wife of a high-profile diver, who was the mover and shaker in the family and would phone around to see if it was a 'diving' day. She would then practically launch his boat and send him off to work. He'd fall flat on his back in the pub later on, either bent or drunk or whatever. There's also the story about the diver who would tow his Haines Hunter boat and truck around supermarkets trying to impress the women with big talk about abalone diving: 'He thought he was somebody. But an abalone boat only looks impressive to another ab diver, otherwise it's just another piece of machinery.'

It's no secret that an investigation was held into the record-keeping in the Zone in the 1990s, but even this didn't deter the divers from playing the system. Harry also tells the yarn about the diver who said, 'This guy "C. Ash" — he must be wealthy.'

But Harry acknowledges that it's a tough job for the divers. 'These guys spend a lot of time underwater on their own as an occupation, and this impacts upon how they relate to others, including their deckhand. If the sea is rough the long-suffering deckie is on board and getting tossed around, while the diver is getting tossed around on the bottom. The diver can come up pretty cranky.'

Harry has mixed views about the future of the fishery and maintains the biggest pressure will come from urbanisation, development and pollution. 'This is a boutique fishery where we mostly export the product to make money rather than consume it. If we don't feed anyone abalone in the future, then it's not going to be culturally devastating, but it will be economically devastating for what will be seen as a relatively small number of people.'

5. Early licence owners

Jurgen (Yogi) Braun

It was gold-rush fever

One of the original divers in the Western Zone, Jurgen Braun was born and educated in Germany before migrating to Australia in 1965. On the migrant ship he met a German who lived in Australia and was returning from a holiday in Germany. He invited Jurgen to go diving with him when they got to Australia. But first Jurgen had to see someone in Melbourne who had arranged a job interview. He told the person that he'd been offered a job abalone diving. The response was that he was mad and that a shark would probably kill him in the first week. Jurgen naively asked, 'What's a shark?'

Jurgen was an electrician by trade and left Germany to avoid conscription. 'I came out as a single man. I met my wife Jan in Portland about four years after I arrived.' Jurgen explains that he got his nickname 'Yogi' because people were unable to pronounce his name, so he'd say, 'Just call me Yogi.'

Jurgen's diving days began in Eden on *The Valiant*, a 50-foot (16-m) ex-cray boat owned by Tony Jones. He was one of six divers on board. A tender boat would drop them on the reef where they wanted to work. It was all scuba work. The tender would bring refilled air bottles for those who had two tanks. 'I only had the one tank,' Jurgen says. I couldn't afford two, so I had to return with the tender and my catch to the mother boat and then refill my tank.

'On *The Valiant* we were a bunch of naïve guys who had no idea what to do regarding diving,' Jurgen says. In those days, because they used tanks, they had to set up a compressor on the boat to fill them up. The only suitable place that was out of the salt environment was in the engine room. On the first day diving down at Gabo Island, after about an hour or two everybody had massive headaches. An hour after they came out of the water the headache subsided, but the next day they got it back again. They had no idea that the compressor had to be outside. 'There were no regulations, nobody knew much about it so you couldn't get information from anybody else. That's how "innovative" we were — we nearly killed ourselves!'

Jurgen recalls that, 'When I arrived in Mallacoota in March 1967, there were 12 abalone divers there, and by the end of September there were over 50 divers, so the grounds were being depleted rapidly. That was one of the reasons why Bernie Morton and I thought we'd check out new grounds. We found Portland. We snuck back to Mallacoota, stayed another few days and then in the middle of the night we packed up and left without saying a word to anyone.'

They moved to Portland in December 1967. Initially they fished for crays, but then chose abalone. They liked it because it was a totally different way of fishing. Instead of being on top of the water in the boat they were under the water — and that was far more exciting.

'When we first started diving,' Jurgen says, 'we'd get an adrenaline rush because we were in a totally unusual environment and nobody else was fishing underwater for any other fish. That was the exciting part of it. You have to remember that most of us divers were about 20 years old and we wanted adventure. We wanted something that was different to a normal way of fishing.'

When they were in Mallacoota they'd expect to get 500 lb (226 kg) of meat, but the 'gun' divers like Rickie Hale would get 1000 lb (552 kg) regularly. 'At Portland you could get 1000 lb of meat virtually at any site along the coast up to the end of the 1960s.

'We earned good money, but it's amazing how much greed comes into it. I remember a time when we had a meeting when the price was about $35 a kg. If you caught 20 tonnes you had a bloody good income. Everybody said if we can stabilise the price at this, we'll be set for life. It was gold-rush fever.'

He remembers that the divers had plenty of cash to splash around and this attracted the local girls, causing the local boys to be jealous. 'All the mums and dads told their daughters to keep away from the ab divers.'

Divers need to be clever at solving problems. When Jurgen suffered back pain, he designed his own weight system to relieve the pressure on his lower back. He started with a hollow plastic backpack used for air dive tanks and filled it with shot pellets. He attached two straps to it that allowed him to place weights around his thighs. It had no weights around the small of his back and had a quick release. He carried a snorkel to use in the event of an emergency.

When one of the divers mounted their compressor on a wooden plank on the boat, they all copied that. It meant the compressor wasn't sitting on the deck taking up room. Jurgen says that winches came on the scene quite late. 'I remember being under the bag lifting it while the deckie leaned over and pulled and rolled the bag over the side of the boat. They were 50-kg bags. Often the deckie would simply hook the bag and lean over and take out the fish while it was still in the water. When it was light enough he would lean over and roll it over the side.'

Every diver had his own favourite spots but, as Jurgen explains, 'You would never tell another diver if you had a big catch, or where you got it.' At Port Fairy Jurgen would check

out the patches at the mouth of the Moyne and the Lighthouse reef break and he would drive to the southwest passage and look at it with everyone else. At Portland he would drive to the passage where there's a reef that they used to gauge where they could work. 'If one boat went out, most would follow.'

Over the years, Jurgen worked his own

way. 'I would go out with others, work with them for a short time at the same place and then shift to where I really wanted to work. I would always leave before any of them. My catches were always less so I wouldn't attract attention. But then I always only worked for four hours. And nobody would know where I had worked because I was home and weighed in and washed down before they were thinking of coming home. I tried to keep my spots secret and I would try and protect the locations by being very cagey.'

Jurgen sold his licence in 2002.

He only eats abalone when he has people over and they want to taste it. And he wouldn't eat it out of a can. He likes it when it's fresh but admits it's an effort to get. If his visitors want to try abalone he'll go out and snorkel for some. 'You have to tenderise it and slice it,' he says. 'I use one of Jan's old stockings and get a brick and give it a whack. The stocking is just to hold it together. Or you can cut it into slices and tenderise each slice. But using the stocking method, with one whack of the brick the whole abalone is tenderised. Cut it into slices, put it into the frying pan and that's it.'

Andrew Coffey
It was luck ... all luck, pure and simple

When asked about the extraordinary life she and her husband Andrew have had in the abalone industry, Valmai Coffey says it was nothing more than luck.

It began in the early 1960s when Andrew was working at *The Standard* newspaper in Warrnambool as a print operator and went diving for shipwrecks on weekends. He saw lots of abalone when he was diving but didn't know if there was a market for them. He began getting a few abalone and Valmai would drive the trailer-load to a fish processor in Geelong on Mondays while Andrew was at work. There wasn't a lot of money in abalone at the time and he continued working at the newspaper until eventually the abalone took over. 'We were only diving at the weekends, but we did it right — we all had a licence.'

They were there at the beginning of Sou'west Seafoods, or Western Divers Co-operative as it was then called. Andrew explains how it all started when Len McCall rang him and said, 'Look, you're working out of Warrnambool so how about joining up with us at Port Fairy and we'll form a co-operative of divers?' The Co-operative

Clockwise from top: Andrew Coffey getting dressed for work. *Photo source*: Murray Thiele. Octopus eating abalone. *Photo source:* David Forbes. Andrew Coffey (centre) packing product. Andrew Coffey with his abalone tool. *Photo source:* Len McCall

worked well for a time and they were selling the abalone to Fishbrook fish processors in Port Fairy. When the wharf building came up for sale, Len suggested the Co-operative buy it.

After purchasing the Fishbrook building, they started buying abalone from the divers at Portland, but found that rival processors Safcol would always pay more. 'If we were paying $2 a kg,' Andrew says, 'then Safcol would pay $2.50. We'd go to that price, then they'd go a bit higher. They were trying to shut us down. Len's pretty smart. He said, "Let's take them on".' A meeting of all the divers in the Zone was called, and they were asked if they'd be interested in forming a co-operative. 'Len said if we could gather the majority of the divers, we might just happen to cause some trouble to Safcol, instead of them causing us trouble. About 90 per cent joined the Co-operative. So eventually, Safcol closed at Portland.'

Getting the product delivered could sometimes be a problem. One of the legendary stories was about a weekend when Len McCall was going away and arranged for John O'Meara to drive the truckload of abalone to Blackney's fish processing in Geelong. He had a couple of blokes willing to help load the abalone onto the truck and was ready to go early Sunday morning. When Andrew rang John on the Sunday morning and asked, 'Has it gone yet?', John explained that there was some trouble with the truck, but he was going to try to borrow Dick Cullenward's truck. 'Okay,' said Andrew, 'but it's got to be at Blackney's at 11 am.' John said that was fine and he would be on his way soon. Later in the morning Andrew rang John's house, and John's wife Maureen said he was a bit late leaving but was on his way.

Not long after that, Andrew got a call from Maureen, saying that John was in Colac Hospital. 'What the hell's he doing in hospital?' asked Andrew. John had had a 'bit of a turn', Maureen explained. Apparently John had nervous blackouts possibly related to diving. Andrew asked where the abalone was and Maureen told him that John thought he'd left the truck on the side of the road at the Koala Motel at Stoney Rises.

Andrew and Valmai drove to Stoney Rises but the truck wasn't there. The motel manager told them there had been a truck parked out the front for a couple of hours, but a man had come along in a station wagon, looked in the truck, obviously found the keys and took off in it. Andrew thought 'Christ, someone's pinched the bloody truck!' Then he realised Dick was driving to Geelong that day to take part in a volleyball championship and had probably seen the truck on the side of the road and decided to drive it to Geelong himself.

Andrew and Valmai went to the Colac Hospital and found that John seemed to have recovered so the three of them drove on to Geelong in the late afternoon. When they reached Mount Moriac they found the truck abandoned on the side of the road. The tread had come off one of the tyres and Dick must have left it there, knowing Andrew would be along soon to fix it. Valmai went off to find a telephone and called Blackney's to explain the situation. The manager wasn't happy but said he'd send his mechanic with a new wheel. Valmai drove back to the abandoned truck and drove off the side of the road and down a drain, tipping her car over. When the mechanic arrived to change the wheel, he pulled Valmai's car out of the ditch.

Andrew suggested that he should drive the truck now, but John was adamant that he would drive. Not wanting to upset John and cause another 'turn', Andrew reluctantly agreed. They continued down the highway to Geelong with Valmai following in the car, when Andrew noticed she was flashing the lights on and off. Wondering what she had noticed, he then smelled something burning and realised the truck was on fire. When they pulled up, Andrew saw that John had left the hand brake on, causing the tail shaft to heat up. It was red hot and had caught fire.

Andrew began pulling the bins of abalone off the truck. John was of little help. He'd gone completely off the rails and jumped over the fence into a paddock shouting, 'It's gonna blow up. We'll all be killed.' Meanwhile, Valmai tried to pull Andrew away from the burning vehicle. 'I got a couple of bins off and thought, "Bugger it, I've done everything I possibly can to fix this," and stood back.' He noticed that it was only the tail shaft that was on fire so he knew it wasn't going to blow up and managed to stop the flames with a rag. He put the bins back on the truck and, as John had calmed down, they continued on to Geelong. 'We got there at night. All up it had taken the whole day. Any longer and we would have lost the lot. We still got paid, but not as much. I told John to take the truck back to Port Fairy and

Valmai and I went out to dinner in Geelong. It was one headache after another.'

Andrew always had a dog on the boat when he was diving, a Dalmatian. On one occasion he had just put on a new deckhand and it was probably the first time he'd been out on the boat. The sea was pretty rough and they launched off the beach at Killarney. While Andrew was under the water chiselling abalone off the sea floor, he noticed he was getting pulled backwards through the water. He couldn't work out what was going on and thought he'd better go up and have a look. 'Up I go and I see the dog's swimming around. I thought, "What the hell … the dog's never fallen off the boat before." Then I look around and think, "Where's the boat?" Boats normally don't sink completely and just the bow was sticking out of the water with the deckhand hanging on to it.'

Apparently the deckhand had stacked all the abalone at the back of the boat instead of starting at the bow and then stacking backwards. A big wave came over and the boat went down backwards, so the dog was swimming frantically while the deckie was hanging on for dear life. Fortunately Lou's boat was nearby and he rescued the dog and then towed the boat back to the shallows. To add insult to injury, the deckhand twisted his thumb and was off work for six weeks. He'd only just started work and they had to pay him six weeks' wages.

Most divers had close encounters with sharks but, according to Andrew, 'You could count on one hand the number of sharks that would be capable of biting you in half.' On one occasion he was diving at Killarney with Murray Thiele. The divers usually wear red gloves, because once they get deep, the red changes to black: 'As soon as the colour starts to change you know that you're in deep,' says Andrew. 'But Murray was wearing purple gloves, so I'm diving away when I see this bloody purple thing coming out of the weeds and I think, "What the hell's that?" I'm always an act-first-ask-questions-later person and so went *whack*. I hit it and nearly broke his arm!'

Andrew and Valmai still live in Warrnambool when they're not travelling.

Rod Crowther
It's a privilege to be in the water
When Rod Crowther started diving full time in 1984, he was among the first of the new wave of divers in the abalone industry.

Born and bred in Portland, he knew that he wanted to be an abalone diver from an early age. He'd been around abalone divers for most of his life and was a deckhand during school holidays for his cousin Bob Ussher. Rod was only 15 when he approached his father for a loan of $200 to set himself up as a diver. But his father said, 'The hell I will. You'll finish your schooling and get a trade or qualification behind you.' So that's what Rod did.

Rod trained as a telecommunications technician, working around the western half of the state for nine years. He resigned from that job soon after getting married, bought some land and built a plant nursery on it where he worked as a professional nurseryman for 14 years. But all the while he was waiting to buy an abalone licence. 'I'd been diving since I was about 12 and was lucky enough to turn my hobby into a profession. I tried to get into the industry for many years prior to 1984. I'd made written and vocal representation to the Fisheries Department to open up the licencing process.'

Before 1984 licences had been frozen and when the Fisheries (Abalone Licences)

Act was enabled to permit transfer, Rod was the first to obtain one. No one had been able to buy in and anyone who wanted to get out was able to surrender their licence for no financial gain. But no one was surrendering their licences. They were still working them. Rod explains, 'Fisheries decided in their wisdom that in order to protect the resource they would allow two divers to sell out to one buyer. So I had to buy two licences to get in. It was called "consolidation" but the shortfall was that one diver coming in was working harder than the two that got out. So he was actually catching more fish than the guys who sold out.'

Rod bought Bernie Morton's licence, after negotiating with the executors of his estate. Bernie had tragically committed suicide six months before the legislation was passed. Rod bought the other licence from Gary 'Ten Bins' Watson. All up, he paid $185,000 for the licence. A couple of divers said he'd paid too much, but luck was on his side, as prime minister Bob Hawke had deregulated the dollar. Just before he bought his licence the abalone beach price was $3.80 and then dropped to $3.20. After Rod bought his licence it went up to $3.40, then $3.80, and

the price just kept going up. All the while, the Australian dollar was devaluing, making the product that was traded in US dollars or Japanese yen just keep on appreciating.

Rod dived out of Portland where, he says, on sunny days when there's not much wind and the water's clear it's a privilege to be out there. But there were many days when they had problems such as with winches or compressors and they'd have to turn around and go back home. 'I enjoyed coming home on windy days with the swell behind pushing from the back. You're riding home on the waves and it's

raining, but you're inside the cabin where it's warm and just enjoying the ocean.'

On some days when the weather turns, the weather gets nasty at Cape Bridgewater. As Rod explains, 'You'd be tucking into the leeward side of the Cape, finishing a bit of fishing there and the wind's picking up strength, but you don't really notice it until you're packing up to go home. Going past Cape Nelson Lighthouse and having huge green rollers just pushing you down into the trough, having the boat skewing and all the gear flying everywhere and having to regain control of the boat and turn it back into the swell, waiting until that set of swells has gone through before you can continue towards the port. That sort of stuff puts the wind up you a bit.'

Rod's first deckhand was John Davies, who was more of a British landlubber than a sailor, having never been to sea before. He was working at the plant nursery when Rod asked him to be his deckhand. On his first day's diving, Rod nearly sunk the boat and John wondered what on earth he'd got himself into. John had been working in the Northern Territory mines, but he thought that was too dangerous. He soon realised the dangers of the ocean.

Another time they were towing a boat, a tinny with around 300 kg of abalone, from Murrell's Beach to the top of the cliff when the car began spinning on the edge of the cliff line. John immediately leaped out, swearing in his northern England accent, and left Rod to it. 'I had to reset the vehicle and reposition it so it had proper grip going up. Finally we made it to the top and it was only then that John decided he was going to get back in because we were safely away from the cliff edge and I don't blame him.' Rod

says he's been very fortunate in his choice of deckhands, starting with John Davies, his brother-in-law Gary Bartle and most recently Andrew Beauglehole.

The divers only work when the weather is suitable so they have to go out when they can. 'Every morning we'd go out to The Gates, at the old quarry at Portland, and we'd check out the sea from there before deciding whether or not we'd work. There'd be several divers standing there talking about this and that, looking at the swell and waiting for someone to make a decision to go. As soon as someone went, everyone went. But you'd try to talk them out of it, telling them that it's not worth going out. It was part of the game, I guess.'

Technology has changed the way the men now dive. These days the divers can check the weather forecast on the internet and see what the swells are at any point in time. They can plan to take days off, which they couldn't do before, because they had to physically go out and check the swell every day. No one wanted to lose a workday, particularly in the days before the quota. 'You'd always be afraid someone was going to get your spot. You'd probably have only half-finished it from the previous run and you couldn't lose a day's income — and it was a substantial day's income. But now that we've got the technology where we can see what the swell is doing for the next few days, you can actually take some days off. You don't have to do the drive out to The Gates at the quarry to look at the swell any more. A bit of a pity really.'

Although the divers are all friends, they are competitive. Rod explains that if he saw someone was working a particular spot and hadn't moved so was clearly having a good day, the next day he tried to beat them out there, drop in on it, look around and decide whether or not they'd finished it. 'That was the only time you'd be trying to tread on someone else's toes,' he says.

The divers usually work within sight of one another. 'Generally, if the weather is in a certain sector and the swell's a certain height, you'd say it's a Blow Hole day or it's a Whites day or a Cape Nelson day, but not a Bridgewater day. So you'd end up having five or six boats in a row because that would be the best spot suitable for the day's diving.' Often there were four or five boats within several hundred metres. When one moved to the next spot, they'd leapfrog one another down the area. 'There was one day in Prop Bay at the Island last year when I think every licenced diver was in the one bay at the same time. It's the first time I've seen that.'

Rod says it was fun racing other boats out to the fishing grounds. 'Two or three boats launching at the same time, getting in the water, heading off and trying to guess where the others were going to go and what spot they're going to. There'd be two or three boats gathered, five boats even, in the one spot trying to talk each other out of diving and then all racing each other home. In one instance we got so close that the front of my boat landed on the engine of the boat in front of me. Didn't do any real damage fortunately, just superficial damage to the engine cover, but you'd think, "That was a bit close", and the next time you wouldn't be quite as game to do that again.'

Since 2001, Rod has been involved in building an abalone farm in Tasmania. Before that, he was chairman of Sou'west Seafoods and on its board for many years. The concept of abalone farming was still very new in 2001 and Sou'west began to investigate whether they should invest in an acquaculture farm. The board ended up getting half the investors from Sou'west to invest in Southern Ocean Mariculture at Port Fairy. Rod was one of those who started that process and he did much of the background work needed for the farm.

Rod Crowther sold his licence in 2011.

Rod cooks abalone by slicing it thinly and frying it very fast with garlic and lemon, or ginger and lemon. 'In the early days we used to beat them to break the fibre, cut them into pieces and dip them in egg and breadcrumbs and fry them that way. They're still delicious.' But he says he doesn't go out of his way to eat abalone.

Sandra Downes (nee McCall)

Life was exciting because the diving industry was new to them

Sandra Downes and her then husband Len McCall are acknowledged as pioneers of the abalone industry in the Western Zone. Certainly they were there at the beginning.

They came to Port Fairy in 1967. Before that they had lived in New Zealand for two years and that was where Len developed an interest in sports diving. When they returned to Australia, Len wanted to dive commercially. Someone Sandra worked with said her boyfriend knew about diving and suggested they go to Cape Conran, on Victoria's east coast. There, they talked to people who were diving for abalone and Len came back very excited about it and immediately obtained a fishing licence, which in 1966 covered abalone diving as well as fishing.

Sandra and Len moved to Phillip Island, where he dived for a while. Sandra recalls those first few months as being very 'raw' and having very little money. 'We spent a month sitting on the beach waiting for the weather to clear up.' They met some interesting characters though: 'Bunch of bums actually,' says Sandra. 'They partied hard all night and tried to dive the next day.'

Life was exciting because the diving industry was new to them. It was new to everyone. They invested all they had in a boat — a Caporn made of wood and fibreglass, as well as a small truck. I think we had about $100 in the bank and that was it — we bought a boat.' They later traded the Caporn for an 18-foot (6-m) Swiftcraft, *The Haliotus* (named after the abalone genus).

After six months on Phillip Island, Len decided he wanted to go to Port Fairy. In those days the abalone divers ran up and down the coast because there were no zoning restrictions. It was open all the way along the coast and they could dive wherever they wanted. Len had heard about Portland divers getting big catches of abalone, although Sandra says he could have got big catches at Phillip Island. When they arrived in the district Andrew Coffey and Murray Thiele, who were already working there, tried to discourage them.

Others came: Dick Cullenward, Harry Bishop and Denis Carmody plus a lot of camp followers, young women who were attracted to the lifestyle. The divers were mainly single men, who partied hard until they needed more money. Sandra says with a laugh, 'There was more diving done on shore with the girls they supposedly took out to shell for them!' It was a good ploy to get girls because, 'Some were nice looking blokes and they sounded exciting, you know, being divers.'

Sandra only went out on the boat with Len when he couldn't get anyone else. Len would shout at her because she didn't handle rough seas well and got seasick. And as a tiny woman, she wasn't really strong enough to work on the boat. It was her job to wait for them on the wharf because they only had one car.

'One day I was out on the boat shelling for Len at Phillip Island, and the next thing I know Len shouts out, "Start the motor!" I started the motor and the bloody thing blew up. Len said, "What did you do?" Well, I hadn't done anything. Anyway we were stranded until we saw another guy in his abalone boat. We waved frantically and he towed us back in. That's when we traded in that boat for the Swiftcraft and got into a lot of debt just to keep going.'

She says that sometimes the boat was so full of abalone they had to come slowly up the river and the edge of the boat would be level with the wharf. That would be about 1000 kg. They used to put about 50 kg in a bin, but now they keep the number down as it makes for a better product. There's a lot more emphasis today on looking after the abalone. Sandra says, 'In those days, you'd shuck them at sea, throw them in the bin and bring them in. We didn't have the Co-op here then. We didn't have anything here at all.'

Sandra wanted to get a job but Len needed her to help drive the truckloads of abalone to Melbourne because he would be worn out after diving for several days straight. The product suffered as well, because of poor processing. It was kept in bins, salted and often left for three days in a truck on the wharf with no refrigeration.

When the Western District Divers Co-op was set up in 1980 it was Len's job to find the best buyer for the product. He established a relationship with Russell Crayfish and they sent a truck down to Port Fairy every day to pick up the Co-op's bins of crayfish and abalone. That made things easier for the divers because they didn't have to do the running back and forth any more, and they had a better quality product, although at the time the processors weren't too worried about quality because Japan took most of it.

Sandra and Len were settled in Port Fairy and decided to stay there. Portland divers stayed in Portland and the Port Fairy divers, with the Co-op, stayed in Port Fairy. But once the Co-op was working well they needed more product so they approached the Portland divers with a view to them buying in. The Port Fairy wives sold their shares to the Portland divers. Sandra says, 'This was because we needed more divers and more product.' Most of the people who bought in at that time are still there, except for Dick Cullenward. When the Co-op's directors decided to turn it into a company, Sou'west Seafoods, and expand into new premises at Awabi Court, Sou'west would can their abalone and send it to Japan. Then, over the years, the Japanese developed their own farms, but they still wanted wild abalone. As Sandra explains, 'Australia and New Zealand are the only places where you can get wild abalone now. South Africa was the other one, but they've closed that down. In America, they had a disease go through, withering foot syndrome, so they haven't got any. They still have the big abalone, the Californian Red, but not enough commercially.'

After living in Port Fairy for a few years, Len decided to try King Island, having heard there were lots of abalone there. After several weeks on the Island, Len phoned Sandra and complained about the lack of abalone and even worse, the lack of social life on the Island. He decided to come back home. He told Sandra he was filling the boat with fuel and would be home the next day. He left King Island early in the morning and arrived at Port Campbell in the dark about seven o'clock that night. He said his behind was so sore from banging up and down in the big seas that all he could think of was getting off the boat.

Sandra tells the story about the men going missing. 'One day we were looking for the men. It was a dead flat day and they were all supposed to be out working but no one could find them. John O'Meara's wife Maureen and I went to Marty Hearne's pub for a counter meal and found the boats pulled up out the back of the pub. When the divers saw us sitting there they said, "Why aren't you two at home?" And Maureen said, "It's lackey's night off!" We thought if they could have the day off, so could we. But I must admit they didn't do that too often.'

Finding and keeping a good deckhand was always a problem. Sandra recalls, 'One time they were short of a sheller. I don't know where they got this guy from but he said he was good. They took him out and he got terribly seasick. Len was on a good patch and the other deckie John was pulling on the rope to tell him to come up. Len was a bit cranky but he came up and John said, "I think we'd better get him back." Len said, "But I'm on a good patch." John replied, "I think we'd better get this guy back, he looks like he's going to die!" Len wasn't happy about it but they brought him in, rolled him on the wharf, sat him up against a pole and went back out. Len said the guy wasn't there when they got back some hours later and assumed he got home okay. We never heard from him again.'

Andrew Coffey used to drive Len mad, Sandra says. Len would ring Andrew to see if he was going out diving and Andrew would say, 'No, it's a bit rough, I don't think I'll go.' Then later in the afternoon Len would see Andrew's boat out there. And Sandra would say: 'I don't know why you take any notice of him, he does it every time.' Sandra thinks Andrew probably didn't get much abalone but he would go out anyway. 'Those things used to happen all the time. They wouldn't change deckhands because they didn't want someone else's deckhand finding out their good patches. But they never really got to fisticuffs or anything like that — they all seemed to get along together pretty well.

Sandra and Len have an equal share in a licence. She was about to sell her half when the virus broke out. She thought that as she was getting older she'd retire and get out of the diving side of the industry. 'As I'm also chairman of the board of Sou'west, I wanted to distance myself from it, but that wasn't to be. So now I'm still there with my half licence.'

Bob Ussher
We're in their territory

Bob Ussher is a man larger than life in both character and stature. He reckons he's had the bends that many times it's not funny, but thought it was probably about 30. He also enjoys telling the stories.

'Fear is being unable to breathe when the ocean surface with its life-giving air is a long 25 metres above.' A local Portland newspaper made that dramatic statement after one of Bob's brushes with death when he contracted the bends and was airlifted to Adelaide

Hospital's decompression unit in the late 1980s. Bob was 42 at the time and had been filling baskets with abalone for two hours when he began to have difficulty breathing. He thought there was a kink in his air hose, so he gave it a couple of tugs. Looking up through the clear water, he could see abalone shells drifting down from his 7-metre fibreglass dive boat, anchored near Portland. Unknown to him, a winch on the boat had got out of control and stopped the compressor pumping to the air hose. 'Once something like that happens it's just gone. You have one breath and the next breath it's gone.'

Bob knew his wetsuit had buoyancy and that when he dropped his weight belt that he would surface — and fast. All he was thinking of was getting to the surface as quickly as possible. He knew he was going up too fast and probably blacked out. The next thing he knew he was on the boat. After about five minutes his shoulder began to ache and he recognised the familiar niggle of the bends. His deckhand tried to radio for help but couldn't get through, so they headed for the Portland wharf. When they arrived, Bob could not move his arms or legs and was almost blind. It was agreed he should go

Clockwise from top: Bob Ussher
unloading his boat at Safcol, Portland.
Bob Ussher with his 300-kg catch at
Nelson. Bob Ussher's first boat, with
deckhand Fred Wolf. *Photo source:*
Bob Ussher

to the nearest major hyperbaric chamber at Adelaide.

A plane was sent from the National Safety Council base at Morwell, Victoria, and Bob was loaded into a portable decompression chamber. He could only just distinguish between light and dark. His next clear memory was of Royal Adelaide Hospital. According to Dr Des Gorman, the co-ordinator of the Diver Emergency Service based at the hospital, Bob had a gas embolism and gas bubbles in his blood, but because he'd been at depth for so long the gas embolism gave him decompression sickness as well. It took about an hour of treatment for the pain to ease and he started to feel comfortable. He was in hospital for seven days and was taken into the decompression chamber each day for three hours. Against medical advice, Bob returned to diving but admits he was 'just about dead with this one'.

Bob loved diving and loved the water. He was always the first diver out in the morning and the last diver back at night. The longest he spent under the water was about eight and a half hours. He loved diving deep.

He first started diving for abalone in Sydney in 1962 when he was a rigger on the Opera House making £19 a week. He and a mate went out diving one weekend and made £40. They looked at each other: 'Why are we working six days a week when we could be diving?'

They worked their way down the coast and got to Mallacoota in about 1964, where they found 'the biggest bonanza of abs. There were abs everywhere. There were only six of us working then, including Phil Sawyer, Derek Fieguth and Dick Cullenward.'

In 1966 a newspaper got hold of the story that abalone divers were making £400 a day. 'Then what could float or anyone who could swim hit Mallacoota.' One day Bob counted 160 boats going out. They took whatever abalone they could find, from the very small to the large, and within one year the fishery was completely wiped out. When they first started diving they were getting 700 to 800 kg a day. By the time they'd fished it out, they were only getting 30 kg. 'So we packed our bags and moved south, looking for new ground.'

They got to Apollo Bay but didn't like it there. They found good abalone around Port Phillip Bay but Bob didn't like city life, so they made their way to Portland. 'There were only eight divers there when we arrived in 1966. One was Gary "Ten Bins" Watson, who has since died. There was "Black Flash", a Kiwi and a top diver; Florian, the Frenchman, who used to say, "I see submarine without portholes!" — he meant a great white shark; Dick Kelly, who has also passed away; Derek Fieguth; "J.C." [Phil Sawyer]; Bernie Morton, who took his own life, and Jurgen "Yogi" Braun. Then there were those Port Fairy fellows.'

They arrived in Portland at Christmas 1966 and not long after the 'big fleet' from Mallacoota arrived and cleaned Portland out. So they packed their bags once again and went to South Australia, where there were only about 50 or 60 divers. 'We went to the South Australian government and told them what was happening in Victoria. So they said they would licence all the abalone divers in South Australia. We had to be in Adelaide on a certain date. All our names went into a hat and they pulled out 150 names. I missed out, so we came back to Portland.'

A fiercely independent man, Bob reckoned the Port Fairy guys used to call him 'the outsider' because he would never join their groups. 'They tried to ban me, but the more they tried, the more I was determined not to join them. I was a good diver. I used to catch anything up to 60 to 70 tonnes a year. Then they brought the quota in. They wanted everyone on a level playing field. They were jealous of me. One year they agreed not to dive for three months and for those three months I had the whole ocean to myself. I didn't see that as saving the resource as there were stacks of fish out there.

'I never played dirty tricks on the Port Fairy guys. They came up to me one day and said, "Ussher, if you don't start slowing down, we'll burn your boat." I said, "Go ahead, it's insured!"

'My biggest catch was 1446 kg shucked [4338 kg unshucked]. That was the year they brought the quota in. I knew it was coming so I worked my butt off. On average we only worked about 60 days a year and that year I worked 94. I worked in some seas that were higher than a two-storey house.'

Bob was close to Phil Sawyer (JC), who was also considered an outsider. 'He wouldn't take any nonsense from the other divers. They treated Phil and me the same way. They'd threaten to sink his boat and cut his diving hose like they did to me. But the more they made it hard for us, the more determined we were to stay the same. We were always together and looked after each other. If he ever had trouble with his boat he'd come and work in my boat, and if I had trouble with my boat I'd work on his.'

Bob loves recounting shark stories. One day he dived in at Lady Julia Percy Island and looked up to see the boat going backwards. He'd landed on a white pointer. 'Until this day I don't know who got the biggest scare. I don't know how I did it. I came back up and didn't even touch the side of the boat.

'I had a theory that I always jumped in about 5 or 10 feet [1.5–3 m] of water and swam out along the bottom, but this day I was sitting there and my deckie said, "You'd better not go in, there's two whites out there." We sat and watched for about half an hour and in they came. God, did they go fast. Then suddenly they took a seal in about 5 feet of water. That buggered my theory up.'

Telling another shark story, Bob explained that if it weren't for a full bag of abalone, he wouldn't be here today. He was working in The Passage and already had about three bags. He was coming up on his last bag when he saw a great white coming towards him. As he could see the shark coming, he held the bag out and the shark took the bag. As it swung around, the shark hit Bob with its tail and sent him flying through the water. Bob went straight to the bottom and curled up in a cave. The shark stopped about 20 feet [6 m] away and watched him. Bob grabbed hold of the hose and gave the deckie four tugs, which is the sign there's a shark, and he brought the boat over the top of Bob's air bubbles. The deckie revved the motor, hoping that would scare the shark off, but it didn't budge. 'I reckon I was there for about half an hour, and I tell you I was getting ready to dirty my pants. Then it moved off. I waited a bit longer and then went straight up. I wasn't worried about the bends that day. I was a bit shaky when I got into the boat. My deckie said I was as white as a ghost. I reckon he would be too if he was down there. We went to another spot but I only lasted about 10 minutes before going home. The next day I had to really force myself to get back in. Luckily, that was when they invented the shark pod. I've used that ever since and if it wasn't for that shark pod I would have given diving away.'

Bob wouldn't change anything about his life. He didn't make much money in his first ten years of diving, but he loved the lifestyle. 'It's only since the 1980s that we started to make good money out of diving. It's so peaceful down there, plus there's no one to argue with. I couldn't think of a better lifestyle. We used to go down and the seal pups would come around and put their noses on your facemask. People would say, "But what about the sharks?" And I'd say, "Well, we're in their territory." I still think that even today we are safer down there than on the roads. We've been here 50 years and not one shark attack. That's pretty good odds.'

According to David Forbes, Bob was one of the nicest men on the face of the earth. He recalls the first time he had anything to do with Bob, he was doing research and going up The Passage on a terrible day. They came around the corner and right at the spot they were going to dive was Bob's boat *Bertha*. 'Bob must have heard our motors and came up. He was wearing the old-style round mask and he took his regulator out and with half his teeth missing he let out a big, "How ya going?" The steam and condensation was pouring out of his mouth, which it shouldn't because it should have been filtered off before it was pumped down to him. But for the next minute he had steam pouring out of his mouth.'

David reckons Bob used to sleep with oxygen beside his bed and if he felt bad he'd breathe some. 'He's been bent so many times, but would go back down the next day "just to get rid of it".'

David recalls Bob being the best boss he ever had and believes he was the best diver ever in the Zone. Bob took the most risks, he said, but caught the most quota. 'I've heard he was the first to catch a tonne of meat. He told me I was the best diver he'd ever seen and it was unbelievable to get high praise from the person I thought to be the best diver. I'd go and work for him and then send him an invoice. Five days later there would be a cheque in the mail from him. He'd never be chasing me to go to work and he paid well — a great guy.'

Bob passed away on 8 August 2012.

6. Other pioneers

Harry Bishop

Twenty years and 20,000 hours underwater

Harry Bishop is proud of the fact that he was in the first group of divers to arrive in Port Fairy from Mallacoota and Phillip Island. Although he only stayed in the Western Zone for a short time before moving to the Central Zone, he has fond memories of the divers and his time in the town.

Harry was an amateur fisherman and he'd heard from Dennis Carmody, who was then selling fencing wire for BHP, that there was abalone at Port Fairy. He chuckles when he recalls how he told Dennis that abalone were 'horrible bloody rubbish things'. But he soon changed his opinion when he found out just how much the divers were making, which convinced him to give it a go.

John O'Meara had relatives north of Port Fairy and said he was going to Port Fairy because there was abalone there. 'So we hooked our three or four boats up and off we went. It would have been about 1967 or 1968. I can remember Len McCall, John O'Meara, Dennis Carmody and I driving into Port Fairy for the first time. It was late afternoon and we drove out to Lower Armstrong between Port Fairy and Warrnambool. We went for a dive with our masks only to have a look around. The abalone were all lined up on the ledges. We thought, "Gee, this is all right. We'll stay".'

Harry bought his first licence, a normal fishing licence, for $6. At the time he says that even Fisheries didn't know what abalone was and so they simply issued the divers a standard fishing licence. Not long after arriving, Harry recalls, there were a lot of guys diving on weekends and the permanent divers said, 'this is no good'. 'There were about 300 divers at the time and we suggested to Fisheries that they put the licences up to $200. That weeded out about 200 of them, which was good. It gave us a couple of years at Port Fairy with very little competition, although there were quite a few divers working out of Portland.'

They had no money in those days and were only getting 25 cents a kilo, unshelled. 'That made a bin worth about $15. In 1968 that was good money but you just didn't get enough good days. I had a mortgage and would often have to go out in ridiculous conditions just to get enough money to make the house payments. I'd even dive from the shore to get a few hundred kilos to pay a few bills.' He says that even the top divers worked hard to make a living in those early days, and some had difficulty even doing that.

Harry used to be a furrier and made mink coats. He and his wife Lorrie got a bank loan to buy a fur machine and a $2000 boat. As it turned out they returned to the lucrative abalone industry, where they could earn as much in a day as they could making furs for a week.

STRANDED BOAT TOWED TO PORT

Police aided by volunteers searched the beach at Killarney late last night for three men in an 18 ft. boat overdue on a trip from Port Fairy.

But while the search was in progress the men were being towed back to Port Fairy, four hours overdue.

Abalone divers Dennis Carmody and Harry Bishop and abalone sheller Des Miller were unharmed by the experience.

They were stranded at sea after engine trouble developed in the outboard motor of their boat.

The men left Port Fairy yesterday morning on a fishing trip and were expected back in port about 7 p.m.

Police were called in to organise a search of the Killarney beach area when the boat had not returned by 9.30 p.m.

Shortly afterwards Port Fairy professional fisher-man, Mr. Brian Newman left Port Fairy in his vessel the Raemur K.

Newman sighted a flare from the stranded vessel off Killarney and later took in tow back to Port Fairy.

Meanwhile, police and small band of volunteers used a four-wheel drive vehicle and a beach buggy to search along several miles of coastline.

Like most of the other divers, Harry had a shark encounter — although only one. 'One day I was diving in about 15 feet [4.5 m] and I saw this great shadow go over the top of me. It was a nice sunny day and I thought there was a cloud going over until I looked up and saw this great bloody tail that went right over me. I had to pull myself down to the bottom to get out of the way of its tail. It swam around me a couple of times and I thought, '"This is it. I hope it doesn't hurt." I got away from it and back up to the boat. The deckie reckons I shot out of the water, didn't touch the sides and landed in the middle of the boat.'

Harry describes the equipment they used as very basic. 'We were using garden hoses and you can imagine what they were like under 120 lb [54 kg] of pressure. They would blow up like a balloon and go *whoooosh*. We were always fixing stuff, even the compressors. We used Briggs & Stratton to begin with but they rusted out. Then Honda brought out a reliable stationary motor and that changed things.' He admits that they didn't know much about compressors and used to fill them with ordinary oil. 'It would leak and we'd be breathing it in. Someone

told us we'd kill ourselves so we started using vegetable oil, which was like taking a huge dose of castor oil all day.'

While the divers worked hard at their job, it was a different story for the deckhands and shellers in the wild days of the mid 1960s. The men worked hard and it's said some were into smoking marijuana. One story is that the guys used to blow marijuana smoke down the diver's air inlets just for fun. But according to Harry, 'Although one of the well-known divers was into smoking dope in a big way, most of us weren't, strangely enough.'

Harry remembers one local they employed as a last resort when they couldn't get anyone else for the day. He was the town drunk and completely unreliable. Even though he was a hard worker, as soon as he got paid he'd go on the grog and they wouldn't be able to find him. 'One day he decided he was going to end it all. So he went to the wharf and wrapped himself in chain and rope, then picked up a railway line and walked to the end of the pier. That's where they found him next morning — at the end of the pier.'

Harry's wife Lorrie is proud of the newspaper cuttings about Harry and his diving mates Noel Middlecoat and Dennis Carmody. They make fascinating reading. One undated story is very tongue-in-cheek and focuses on Dennis Carmody, describing him as 'The King of the Bay and a ladies' man on the side …' and 'a lean, long and lithe man standing 7 feet 6 inches [2.29 m] tall.' The article is just shy of telling readers how Dennis could leap tall buildings in a single bound, but Harry is quick to point out that the journalist wrote the article in the pub and considerable journalistic licence was taken to embellish the facts. Perhaps it was the liquid refreshments at the time, but Harry confirms Dennis wasn't quite 7 feet 6 inches tall.

Lorrie delights in a small article that looks like it was from what was then the *Sun News Pictorial* (now Melbourne *Herald Sun*): 'Stranded boat towed to port'. Lorrie remembers the night well. 'No one would tell me about it and I thought Harry was at the pub getting drunk.'

Harry explains: 'We were on our way home between The Cutting and Port Fairy about half a mile [800 m] off shore when we broke down. We joined two 100-foot [30.5-m] hoses and attached these to our 200-foot anchor line so we could hit bottom. A storm had come in from the northwest and every time a wave passed we could feel the hoses stretching and we thought, "This is not going to last." Anyway, I went to the front to have a sleep. The others said I was mad, but I said that if we were going to finish up having to swim, then I wanted to be refreshed. So I had a nap.

'At about 11 pm we really started to worry. That night there was a dinner for the local fishermen. They went to the dinner and realised we hadn't arrived. They all came out in their suits and ties in Brian Newman's boat *Raemur K*. We didn't have any flares left, but we had ab bins, so we threw petrol into them and set these alight. Then we pushed them overboard. When the fishermen came and rescued us, they said the sounder had found that our anchor had hooked onto the only piece of rock in the area. We were lucky.'

After leaving the Western Zone, Harry fished for abalone in the Central Zone for 18 years. Lorrie reckons that, based on Harry's returns, he has spent well in excess of 20,000 hours underwater during his career. But after more than 20 years he says, 'I'd had a gutful.'

Harry and Lorrie retired to Broadbeach, Queensland. Lorrie admits that things have now changed for the couple. 'I used to be known as "the wife of Harry Bishop, abalone diver".' Lorrie took up competitive triathlons when they shifted to Queensland and says that Harry is now known as 'Lorrie Bishop's husband'.

Lenny Burton

Wild men o' the west

There aren't many people who can say they have surfed a whale, but Lenny Burton is one of them. They were out on the Greens, an area on the north shore of Portland Harbour in the shallows where the greenlips are, and a whale was swimming underneath them in beautifully clear water. 'He came up with his fluke, his tail, and the water swelled up a bit, making a circle.' They were in the boat going along behind the whale as it came up again and then went down. But this time it didn't come back up. 'The next minute he comes up again but he's right under the boat. Up he came and we were lifted right out of the water — a 19-foot [6.5-m] boat.'

Lenny was behind the wheel looking over the side and Billy Simpson, his deckie, was hanging on to the bow. The boat went sideways and Lenny thought they were going to be hit by the whale's tail. 'I'm looking straight down his blowhole and out comes all this spray. Billy's got one foot over the bow rail, he doesn't know whether to jump off or stay in the boat. Anyway, we just slid off the whale and he went on his way. We surfed the whale! That was exciting.'

Lenny dived for 20 years, starting in 1968. He was in the building game in Melbourne and looking for something different. He had done some amateur diving with his brother and, when he saw an advertisement in *The Age* for abalone diving in partnership with a Dick Kelly, he decided to apply. Dick had the boat and Lenny had a bit of gear.

They went to Blackney's at Geelong and asked them where was the best place to find abalone. Even though all the action was at Portland, they were sent to Torquay because Blackney's was getting too many blacklip and wanted greenlip. 'Everyone had already been to Torquay and fished the place out. We didn't know that so we were the only two divers there.'

It took them about four days to find their first abalone, because they didn't know what they were looking for. Eventually they saw a rock that moved and realised that was an abalone. 'Once we knew what we were looking for, we were right. We'd never seen a greenlip. We'd only seen the blacklip when we were diving at Phillip Island. We only made $20 each the first week so we didn't do too well.'

Then they discovered that all the divers from the other side of Lakes Entrance and Mallacoota were heading to Portland. By the time they got to Portland, there were already about 60 boats there. 'Blackney's only had one big truck and if you were home too late you couldn't sell your abs. You'd have to tip them back out to sea.'

A licence at that time was only $6 but it gradually went up. The price of abalone then was 25 cents a kilo for the meat. One time Lenny lost his weight belt when it fell among the abalone and went to Blackney's in the truck: 'It was worth more than the abs were worth, so I was glad to get that back."

Lenny thinks the people who became abalone divers would have been like those who went to the Wild West or gold prospecting a century and a half ago — fiercely independent, the ones who looked for adventure or something different. 'I wasn't so much like that. I was a bit more stable or settled than a lot of them. It was good money and not a lot of work. When you worked, you worked hard but you still had time to do other things. I established a little farm in that time.'

Lenny remembers Noddy Hill as a colourful character. 'He was a likeable bloke and he loved making things. He was always building something. He didn't have a wife so he had a lot of time to make things. He was always in a boiler suit. At parties he'd nod off after a few beers and then go to sleep, hence the name Noddy. I was never really involved with the parties because my wife was fairly strict with me. When I first started to get into the fun times, she pulled me up fairly quick.' Lenny recalls that the Western Zone has never lost a diver, only two men who were shellers for Noddy.

At one stage the divers wanted policing of the abalone industry and the introduction of abalone sizing. Lenny says Fisheries wouldn't do it so they decided to do something themselves. 'JC, Bob Ussher, Derek Fieguth, Dick Kelly, Bernie Morton, Ron "Wacker" O'Brien, Rick Harris, Tony Jones and myself. We'd have our meetings and we'd all be talking about different things, including undersized abs because we knew there were a couple of people who would consistently pull them. We wanted to conserve our industry so we arrived at a system where two people would be on duty to check other people's boats.'

Lenny thought Phil Sawyer was a good diver, one of the best divers in Portland. 'He could dive at 90 feet [27 m] all day and not get bent, while others like Bob Ussher and Noddy Hill got bent in 30 feet of water.'

Lenny enjoyed his time in the industry, but after 20 years he thought he was ready to try something else. He got out before the licences cost big money. It was during the consolidation of licences and Rod Johnson wanted to get out as well. They sold their licences to Lou Plummer and each received about $100,000.

Lenny cooks abalone on the barbeque. 'I hammer it like hell and cook it for 30 seconds and it's not bad.' He remembers when it was called muttonfish and was only used as bait for crayfish: 'They go mad on it. That's how we caught the crayfish at the end of the day. We'd get one ab and put it upside down where the cray was in a crevice, and the ab has to turn back over and that's very difficult for it. While it's doing that the crayfish is watching. You get around the back so it can't see you and the crayfish will rush out to grab the abalone before it can turn completely over. That's when you grab the cray. The abs are always found in crayfish grounds except when there's flat pebbly ground or sandy ground. The crayfish need a bit of rock cover.'

Dick Cullenward
Port Fairy's Peter Pan

Dick Cullenward was the stuff legends are made of. A champion swimmer in the USA, he was chosen as part of the US water polo team for the 1956 Olympic Games in Melbourne. And if that wasn't enough, he was

His dream meant giving up his naval career and it took him five years to make it happen. When interviewed for this book, Dick recalled how in 1956 there were no dive shops in Sydney but when he returned five years later there were 28.

Dick first settled in Tasmania and opened the first dive shop in the state. He worked six days a week in the business. 'Being a brilliant businessman, it took me almost four years to figure out that on the Sunday I was diving for ab I was earning more money than the other six days in the business. Really dumb!' Then he met some Mallacoota abalone divers who claimed they were getting 500 kg of abalone in a few hours. 'They were getting 20 times what I was getting in Tasmania. So I moved over to the mainland and that's how it all started.'

also a dashing naval fighter pilot. But with all those achievements, Dick chased his dream of opening a dive shop in Australia and diving for abalone.

Dick headed to Marlo in Gippsland and dived for abalone for a while, but he found it was all too new and huge areas hadn't been explored. He then went to Phillip Island and his boat sank. That's where he met Len McCall — he thinks around 1967. 'While I was waiting for my boat, I headed up the coast and down to Port Fairy. I thought I'd be there for a couple of weeks but ended up staying for 27 years.' He established himself in Port Fairy at the very beginning, along with Andrew Coffey and Len McCall.

These were the fledgling days in the Western Zone and the abalone divers were only fishing to sell. Buyers were purchasing their catch and, according to Dick, paying a 'ridiculously low price, then selling it overseas and making a hell of a lot of money. People like Len McCall, John O'Meara and I were selling our catch to various people. Then we realised this was crazy. We realised we should get all the divers in the area to sell to the same person and that's how the Co-op started. Then the place on the wharf came up. It was fantastic and I thought there'd be no chance we'd get it. But we did. At one stage we were the biggest employer in Port Fairy,

employing maybe 20 people during peak times. It was a big deal for the town.'

Dick recalled a meeting at his home around the time divers were getting 30 cents a lb (0.45 kg). When Phil Sawyer said that he imagined the price would get to $1 per lb, everyone laughed at the thought of making $500 a day.

Murray Thiele believes Dick was probably the best diver in the Zone. He recalls how Dick had a very distinctive twin-hulled yellow boat called the *Double Uggly* and that the other divers used to call Dick and his deckie the uggly double on the *Double Uggly*.

Dick enjoys reminiscing about the colourful characters he knew in the Western Zone — people like Red Quarrell, who dived at Port Campbell. At one stage, Dick said Red leased his licence to his son. One day they were diving and Red's son wanted to go home. Red said, 'No', but his son insisted. 'Red got really pissed off and threw him over the side, sailed his boat back to shore and then drove to Melbourne, where he told Fisheries that he wanted to get his ab licence back. His son had to swim home.' Red Quarrell died in December 2010.

'They were fun days. The beauty of it was that I got my quota and only worked 42 days of the year. I wouldn't go out on a "scratchy" day, but Andrew Coffey, Lou Plummer and those guys would work their butt off in the big waves. It was an absolutely magnificent job on a beautiful day.

'There was this guy who was my sheller. We used to pay them $1 a bin, so if we got 10 or 15 bins, that meant they earned $10 or $15. This guy was terrific but he came up to me one day and said that he had to leave because he couldn't stand the screaming. He said, "I lie there at night and I can hear the abs screaming." What can you say to a guy like that, except to tell him to get off that stuff you're on and you won't hear them.'

Dick told how some would not dive around Lady Julia Percy Island because of the sharks. He remembers a cray fisherman called Russell, who would take a 44-gallon drum and put half or quarter of a sheep on it. 'I'd see the great big white pointer there and I'd say to Russell, "There's a big one out there, probably 16 feet [4.8 m]. He'd go out and set the drums. He'd catch a shark and bring it in, then notify the media. They'd hoist the shark up and take pictures of it and it would be on TV. But eventually they put a stop to doing that.'

Then there's the legendary shark known as Big Ben. Dick reckons that in the early days the cray fishermen thought the shark was 30 feet (9 m) and they would throw stuff for him to eat so they could get a look at him. To get a measurement of Big Ben, one person would stand on the bow of their 47-foot (14.3-m) boat and another on the stern as Big Ben swam by. The first one would shout 'now' as the shark passed and the other would run to the back. 'That's how they reckon he was close to 30 feet,' according to Dick.

'Headlights' was a huge sting-ray that Dick said was as big as a table. 'I'd be down there diving and all of a sudden I'd feel "whoomp, whoomp" and there's Headlights right in front of me. I'd get him an undersized ab and stick it in his mouth. He'd stay right there with me and I'd hear this crunch and he'd chew the whole ab, shell and all. I used to see him maybe 80 per cent of the time I was diving. He'd come back for a few abs at a time. Sometimes he'd be above me and I'd just push him away. He was fun. He was one of the things I really remember.'

Actor Lee Marvin used to visit the Western Zone regularly and go out fishing. 'He was a good guy,' Dick said. 'I had drinks with him a few times.'

Dick's first wife Natalie says that life was good and fast for them in sleepy Port Fairy. They developed a circle of friends around the divers, a cocktail club, the local theatre and Dick's water polo. 'Being married to Dick was absolutely fantastic. It was like being married

to Peter Pan. He was a man who never grew up. Life was never dull.'

Dick's recipe for Californian abalone steaks is original. First you cut the sole out of the abalone — it has to be cut into thick, not thin, strips, and you have to pound it to tenderise it. That's important. Dip each piece in beaten egg and then breadcrumbs, and put about nine pieces into hot oil in a pan. The secret is to then turn in a circle three times and clap your hands. Turn the abalone over and do the same thing again and it's ready. Put butter and fresh lemon onto it and have a good wine or beer with it.

Dick always planned to retire at 60 and move north and this is exactly what he did. He lived in Noosa with his wife Cheryl 'Swifty' Swift for many years and passed away in January 2014.

Derek Fieguth
We were all free spirits

Derek Fieguth was among the first abalone divers in Victoria and in the first wave at Portland. He now lives in Indonesia and sometimes the Northern Territory, and admits he hadn't given abalone a thought in 20 years before being approached to contribute to this book.

Having always lived near the sea, Derek was a keen spear fisherman. When he heard there was abalone at Mallacoota he headed there. He spent time fishing on the New South Wales coast and ate abalone as a meat substitute, never dreaming he'd make a living from what he called 'this unloved desperation food'.

Derek says it was expensive to buy equipment in those days. He went into partnership to buy a boat and outboard motor with four or five other divers and these men became the core of the Mallacoota Divers Association.

After a time they noticed abalone stocks were being depleted in the area and they decided to move to Marlo. Derek stayed for a year and a half. Again stocks depleted and nobody had any idea about the long-term future of the industry. At the time the prices were very low — Derek thinks about one shilling (12 cents) for a lb (0.45 kg) of abalone.

Then the divers moved across the state and on to Port Lincoln in South Australia, where they found the fishing good because the divers were organised. But, as fate would have it, he and the others were ousted from the area amid fears it was going to be fished out.

'I was heading back to Mallacoota but ran out of money and didn't have enough petrol to get past Portland. That's why I stayed there,' he recalls.

When zoning was introduced, Derek had to choose a zone and settled on the Western Zone. 'That year was a drought year,' he says. 'The weather was fantastic and Portland looked beautiful. So I signed up for the Western Zone.'

There were around 10 divers at Portland at that time and the seas were still abundant with abalone compared with other areas of Victoria, New South Wales and South Australia. It didn't take long for the number of divers to increase to between 40 and 50.

The divers around Portland formed two camps: one group in Portland and the other at Cape Bridgewater, where they lived in the caravan park and worked off the beach. According to Derek, the Cape was a short-lived community of perhaps 20 to 25 divers. Because they mostly fished from small 'tinnies' they couldn't go far afield, so they concentrated on the reefs in the sheltered lee of the Cape. After about a year, those areas were depleted of abalone and the divers dispersed. Derek recalls that although there was always competition between the two groups they worked together, especially when they needed to get enough abalone for a truckload.

About the time zoning was introduced in Victoria, the licence fee was raised to $200. The fee drove some divers from the industry while others scattered around the state and one or two became town divers. In later years the divers from Portland would sometimes launch from Bridgewater beach and fish the reefs there, but the caravan park abalone diving community was by then history.

Some divers would only deal in cash in the early days. Derek recalls, 'The divers didn't trust cheques so buyers used to drive from Melbourne carrying enormous amounts of cash, maybe as much as $100,000 on board, and there was no security back then. Of course, this meant the divers had lots of cash to splash about and some didn't necessarily tell the Australian Tax Office about it. At one stage there was a tax investigation into the divers' incomes.'

Derek took on the role of secretary of the Portland Abalone Divers' Association for a while because nobody else wanted to do it. It was a fairly invidious position because, he says, no matter what he did not everybody was happy. He describes the divers as a group of individualistic men who had their own ideas about how the industry should evolve and how to make it better for themselves.

One of the major issues was zoning, because in the early days the whole of Victoria was open and divers could fish all over the state. Derek recalls how the divers from Port Fairy and Portland, who were a long way from anywhere, were looking after their abalone by not taking undersize fish and looking after the fields better than divers in other areas. However, this self-imposed conservation made the waters rich pickings and outside divers raided the stock and took undersize abalone left behind by the local divers. 'This was one of the major reasons local divers pushed the government hard. I represented the divers.'

The other major issue Derek worked on as secretary was the selling of abalone licences. 'We'd all got our licences for practically nothing and when it finally came to selling and getting out of the industry, by then natural attrition had taken care of many of the divers. The original people in the industry felt they'd spent many years developing markets and technology, and lobbying for the many benefits the industry enjoyed. So we figured it was worth something when we sold out.'

As secretary of the Association, Derek sought to lift the profile of abalone divers and raise discussion concerning the issues surrounding the industry. Divers in the Zone were often quoted in newspaper reports, particularly about issues such as poaching, conservation of stock and the effect of floating the Australian dollar on the income of divers.

In 1972 Melbourne's *The Herald* newspaper chronicled the dangers facing divers and Derek was quoted discussing the risks the men took;'... until 12 months ago, divers were putting themselves at risk for the money they could get. They used to dive up to any depth, for any length of time. We were worried about the number of cases of bends resulting from this attitude,' he said. 'They were exceeding the times laid down by the United States Navy for time spent on the bottom by as much as 400 per cent and in a lot of cases apparently getting away with it. But it has been shown now that there are long-term risks, which we were not aware of at the time.' Derek said the first serious case of the bends occurred at Portland in 1968 — the diver was 100 ft (30.5 m) down and was forced to surface rapidly because his compressor stopped and he ran out of air. Derek said the Association wrote to various experts, but no one could tell the abalone men the likely long-term effects of diving under the unique conditions of their industry.

As the Fisheries inspectors did not have a boat to go about their business in the early days, there were occasions when Derek, as secretary of the Portland Abalone Divers Association, and his boat were commandeered by the inspectors. 'I was bearing the brunt of the attack on the guys taking the undersize abalone. In a sense my equipment was used as a lynching tool.'

There were also times when his boat was used to bring in people who had drowned at sea. 'This was hazardous work and I think I was awarded an honour from the Royal Humane Society,' he recalls.

One of the most disturbing rescues Derek was involved in was when Paul (Noddy) Hill's deckhand drowned. 'That was a particularly rough day and we could see the storm coming, so we made a run for home. When I got back into Portland harbour some tourists said they'd been to the top of Cape Nelson and had noticed a boat not far out with the two guys on board waving madly. The weather was deteriorating and my deckie and I went back out but the seas were huge. We got two-thirds of the way to Cape Nelson and I told my deckie that I didn't want to endanger his life because of the hazardous conditions. Just as I made that comment I saw a flash of orange on the horizon and it turned out to be Paul Hill with a life jacket on. I managed to get to him and drag him onto the boat but he was in a state of shock. His boat had been picked up by a wave and thrown against the cliff. His deckie, Freddie Heilien, 'German Freddie', had refused to abandon ship and died as a result. I knew Freddie well. Unfortunately, I was one of the guys who had to drag him out of the water when they found him a week later. Paul lost two deckhands in a few months.'

For Derek, what stands out about his time diving at Portland was how the divers were an eclectic group who came from all walks of life. He says they didn't have much to lose by adopting the free and easy lifestyle of an abalone diver. He recalls there were few constraints on the men's behaviour and it was a lifestyle well suited to their eccentricities. These were independent men, both socially and economically, and their lifestyle meant they never had to curb their proclivities. When you're making $1 million a year this sort of money brings out eccentricities.

'We were all doing okay because we were all totally free spirits. For me it was worth it, to have absolutely no constraints. The other aspect was that we became significant players in the local community. These guys had a lot of money and so they were well respected. If we drove down the street almost everyone in town would wave to us, even though we didn't know all of them. One of the things I became conscious of when I stopped being a diver was that I became a nobody. For most of my life I'd been "Derek Fieguth the abalone diver". I'd done this in such an intense way for such a long time that when I stopped diving I felt like a nobody. It was an interesting experience.'

In the early days Derek recalls how the divers were doing well, but not earning the sums of money they did in the later years. On reflection, he thinks he sold prematurely — a couple of years later people were selling their licences for millions. At the time he was earning around $300,000 and had a simple formula to work out how to disperse his income: $100,000 to him as wages, $100,000 for 'Treasurer Keating' (ie, for taxes) and $100,000 for expenses. He reckons this formula was pretty accurate.

Derek left the industry around 1990 after almost 30 years. By that time, 'The price of our bit of paper had gone up to about a million dollars. So I thought I'd try for another life while I still had some life to live.'

These days he spends his time travelling around Indonesia's 17,000 islands, where he surfs and enjoys the diversity of cultures.

Tony Jones

If a shark is going to bite you, then so be it

An early entrepreneur of the abalone industry, Tony Jones started diving in 1965 at Cape Everard near Mallacoota with a crew of 14 on a boat called *The Commissioner*. Tony had bought the vessel in Melbourne and sailed it to Eden. They were on their second voyage when the boat sank.

The insurance company wouldn't pay the claim and Tony was left virtually broke. He camped at Bastion Point in Mallacoota and there he came across Bob Ussher. Tony had met Bob earlier when he was working as a dogman on the Sydney Harbour Bridge.

When Tony got to Mallacoota he found a group of divers and a lot of rivalry. Chinese restaurants were interested in abalone and Tony worked to establish contracts overseas. Phil Sawyer, Derek Fieguth and Tony went to see the Victorian government to try to get some assistance but were told that they were kidding themselves and that there would never be an abalone industry. 'They showed us the door.'

In 1968 Tony purchased a Haines Hunter and an old car then drove from Sydney to Portland. He stayed at the caravan park in an old army tent. Tony recalls how the abalone industry attracted interesting people from all walks, such as Gunther from Germany. They

Clockwise from top: Launching his boat. Tony Jones diving. *Photo source*: Tony Jones

had nicknames for each other like 'The Gun' or 'The Pistol'. Tony was just known as 'T.J.'

There was always a lot of rivalry among the divers. 'You'd get little jealousies within the industry, I'd go to a spot and look out to see how the weather was and if there was somewhere I could work. I'd go to work. But there were times the other divers would say, "Why do you want to go out today?" But I'd go out and end up getting a good load.'

But most of the time they all got along with one another. Phil Sawyer, Derek Fieguth and Tony played cards when the weather was too rough to go out. Sometimes they'd play for two days straight.

Tony recalls that at one point a group of divers confronted a diver who wasn't observing the size limit: 'His excuse was that he was leaving anyway and he'd take whatever size he wanted.' Tony told the group they should try to reason with the diver rather than yelling at him. He thought they should look at the size of abalone he'd collected and, if they were undersized, they should throw them back.

As Tony says, 'In any industry you have to conserve what you've got. We were the ones who brought in that size limit, not the government. We said we shouldn't be shelling at sea. Instead we should be bringing the abs in unshelled. The Portland Abalone Divers' Association linked up with the Port Fairy guys and agreed to do this.'

Tony dived from 1965 to the 1990s. He has bone necrosis in the shoulders that he attributes to having the bends. He got bent outside the Lighthouse, where the water is over 100 feet (30 m) deep. On that day a big swell came through and his hose was pulled off. 'I had 500 feet of hose out and with the wind behind it, it was pulled off.' He was down 100 feet and had no air. 'So I had to get from 100 feet to the surface with no air. I came up too fast. Even though you try to come up slowly, you can still end up with the bends.' He was taken to the decompression chamber at Prince Henry's Hospital.

Tony was always innovative throughout his diving career. He recalls that in the early group of divers there were a couple of old guys, probably in their 50s, who wore weights around their ankles. They used to walk around the bottom of the sea floor with the weights keeping them down, collecting abalone. 'They did it that way until we realised the best way was to use a hookah because the hookah is a compressor and you've got a reserve tank.'

Tony always had two compressors on his boat, one on each side, and always had two motors. The guys would ask why he needed two compressors and he'd tell them that one day if he was 60 miles (97 km) out and his compressor broke down, then he'd have another in reserve.

Tony explains that a lot of divers didn't want to work Lady Julia Percy Island because of the sharks. He tells the story of a German diver, Gunther, who had arrived in Portland from Mallacoota. 'Gunther was here two days and he had two bad days. He had a boat and trailer and the first day the deckie didn't unhook the boat from the trailer and the whole lot went into the water. The next day Gunther went over to the Island and was in the water about 15 minutes when a shark took a seal right beside him. The seal had been bitten in half and its entrails were hanging out. Gunther went back to his boat and back to shore, where he packed his bags and went back to Mallacoota.'

Tony dived at the Island. 'I'm a bit of a fatalist. If they're going to bite you, then so be it — but the chances are slim.' Sometimes it's a battle out there. He had a couple of favourite spots, which he went to frequently, one spot they named 'Jones' Bay' because he was there so often. But the men used to play tricks on each other. 'I wanted to be the first out and throw my anchor in. The others would see where I was and go on to Bridgewater. But as soon as they'd gone I'd up anchor and take the boat further back. The next day they'd think I'd worked that area but, of course, I hadn't and then I'd get plenty of abs.

'Another trick that isn't so funny was when I had another diver on my boat. I would see where he was working, on a ridge say, and I'd take my regulator out and lay on the bottom. He would come over thinking something had happened to me. There was another thing I'd do when I had other divers on the boat and we were all using the one compressor. I'd lie on the bottom breathing as heavily as I could and force one of the other divers up. By forcing them up, especially if we were on a good patch of abs, I'd get all the air.'

Tony was catching crayfish as well as abalone. 'I used to have really large bags for abalone. If the bag was too big, the boat would tip to one side as it was being winched up. Sometimes it was amazing how many abs you could get in a bag. In certain areas the larger bag meant you could cover more ground without having to get back to the boat. I used to have 400 to 500 feet [120–150 m] of hose line.'

Divers need a good sheller or deckhand, Tony says, and 'Tunner' was probably the best one he had. Tunner's parents were Solomon Islanders and he was born in Rockhampton. 'He was very keen eyed. I'm an asthmatic and sometimes I'd have an attack while I was on the bottom. I'd think the compressor had stopped and the only way to find out was to take the regulator off and hit the button. If there was plenty of air, I knew then I was having an asthma attack. When I got to the surface, if I still couldn't breathe I'd get Tunner to cut my wetsuit top off with scissors so I could breathe more easily. I went through a few wetsuits that way.'

Tony still lives in Portland and is engaged in various business activities.

Ron O'Brien
The youngest of heart
Ron O'Brien was a different sort of man, according to Kim Heaver, his deckhand of 15 years.

Kim recalls Ron being the oldest diver in the Zone at the time, but fondly remembers him being 'the youngest of heart': 'He was one of those guys who never grew up. Even at the end he was still young in his mind and trying new things. He was into martial arts for a while, then he bought motor bikes, and then he was into motor bike racing and doing things blokes half his age would do.'

Born in 1929 in Sydney, Ron trained and worked as a mechanic. He was a keen amateur diver and belonged to a skin diving club in Sydney, which formed the nucleus of the divers who went into the abalone industry. They went to Mallacoota and then on to Tasmania and South Australia — some even going on to Western Australia. Many of them returned to Mallacoota.

Before the Mallacoota divers arrived, there were pockets of local divers in Portland who used a flying fox at Cape Bridgewater to haul the abalone up the cliff. They were Borthwicks meat workers who dived part time. According to Ron, the price they were paid was a lot less than what was paid for the fish in Melbourne.

Ron competed in spearfishing competitions against legendary documentary maker and diver Ron Taylor and he took Valerie Taylor for her very first dive.

He started diving for abalone in Mallacoota in 1960–61, arriving a few months after the first group of divers. The originals were the blokes he used to fish with in the team events at the fishing competitions. He dived using a hookah, rather than free diving. Ron was in Mallacoota for a few months then went on to Portland. Phil Sawyer had been in Portland and then returned to Mallacoota. When he left Ron followed him. Others did too, including Freddie Kurtz and Derek Fieguth. Rick Harris came from Melbourne. Bernie Morton and Jurgen Braun, Bob Ussher and Tony Jones all came from Mallacoota.

Ron had freedived for abalone all the way down the New South Wales coast, working in pairs. He'd chip and his partner would pick the abalone up and put them in a bag suspended under an inner tube. The real changes occurred when they started using

large compressed air cylinders of nitrox, which was safer. This way they could dive deeper and not be concerned about the bends.

At the time the quality of wetsuits varied a lot. Some were good and warm, others were surfboard-like. There were others that were useless in deep water because the neoprene compressed too much and didn't spring back. Ron started with two jumpers stitched together, then got on to seal skin, which was like inner-tube rubber. They could wear the jumpers underneath, making it marginally better. When Ron bought his first American wetsuit, he discovered it was incredibly warm and he started winning spearfishing competitions in this suit. Ron also used an inner suit, like a surfer's steamer, under his wetsuit. Portland's cold water could knock the divers around, but if they wore too much neoprene this could tire them out as well. Ron liked Phil Sawyer's idea of pumping hot water into his wetsuit, which meant he could dive deep all day and never come up. Ron always wondered how Phil never got bent. Konrad Beinssen and Phil worked with a collection cage so that they only had to surface twice a day. They systematically worked The Passage like farmers ploughing a paddock.

In the early days in Portland they'd phone a buyer in Melbourne and he would send a truck to meet them at the wharf. They did it collectively and, because of the size of the truck, they put themselves on a daily limit of 400 lb (180 kg) of meat. Ron recalled that Tony Jones was probably the first to break away and start doing things on his own, including organising his own transport and taking it to Melbourne. Safcol set up in Portland to buy the abalone and eventually the divers set up Sou'west Seafoods because they knew they could organise better deals than they had with the processors.

Ron maintained that it was the divers who instigated the important decisions, not the Victoria Fisheries Department. Ron was secretary of the Dive Association at the time and recalled they were the ones who pushed for zones. Other divers had come to Portland and got 800 lb (363 kg) at Point Danger and the Portland divers tackled them about it, thinking that the only way to stop them.

In an interview with the *Portland Observer*, Ron was asked how long the abalone industry could last in Portland waters. In his opinion, he said, it could last indefinitely. 'There will always be regrowth of the abalone if only a limited number of divers work the region and self-controlled conservation is carried out.' He also considered the greatest danger to the industry was 'nomadic non-conservationists' who could come and harvest indiscriminately. 'These people could kill the industry overnight by fishing the beds out,' he said.

This interview uses some material from an interview with Ron O'Brien by Robert Coffey on 13 June 2003. Ron O'Brien has passed away.

Gary 'Ten Bins' Watson
A Bex and a cup of tea

As a schoolboy athletics champion at Brighton Tech, the young Gary Watson wasn't much interested in schoolwork. He wagged school to go fishing so often his teachers sent notes home complaining that he was never at school.

Gary's widow Bev says his first love was always fishing and diving. He wagged school to work on fishing boats and also did some amateur diving in Port Phillip Bay. Bev says that after leaving school, Gary worked as

a licenced plumber and then later as a real estate agent, but he hated both those jobs. He really wanted to be a professional fisherman — and he got his wish.

Gary started diving for mussels in Port Phillip Bay in 1962. At the time they harvested both abalone and mussels, but often couldn't sell the abalone. He didn't have a boat so he used a tractor tyre with a hessian bag tied on and scuba gear. The catch was sold to various outlets around Victoria, but the abalone was difficult to sell. On more than one occasion they put the bags of abalone on the back of a train going into Melbourne and the bags were picked up at the other end.

Gary went to Mallacoota in about 1964, along with every other abalone diver it seemed, and also to Eden for a short time. He then went to South Australia for an exploratory trip, but settled in Portland for the first time around 1967 before going to Tasmania for about 18 months. 'He would have been four or five years into the industry by then,' Bev explains. They settled in Torquay in the early 1970s before going back in Portland in the 1980s.

They built a house in Portland, then had a motel on 3 acres (1.2 ha) of land right on the beach. Around this time Gary got badly bent and after that he was unable to dive to any depth without getting bent. He decided to get out of the industry. He and Bernie Morton transferred their licences to Rod Crowther in a 'two-for-one' deal.

Gary concentrated on cray fishing and looked after the motel that they had leased to Alcoa. But nature stepped in — in a big way. A huge sea came in and took out most of the beach and the road leading to the motel, creating a 3-metre drop. He and Bev had to get out of the motel owing money to the bank.

They went back to Torquay, where Gary continued cray fishing. On a cold, blustery, winter's day on 29 May 1991, Gary was fishing near Point Addis when his boat was hit by a huge wave and capsized. His 22-year-old son Morgan survived the ordeal by swimming 200 metres to shore. He told police that he last saw his father swimming on his back, but Gary disappeared near rocks at the bottom of a cliff. His body was never found.

Harry Bishop remembers that Gary was very fit and it was hard to imagine him not being able to swim into shore. 'We used to have a laugh and I'd say, "Have you been working, Gary?" And he'd say, "Yeah, I've been getting a few." Well, I looked at his boat on the grass outside and the bloody grass was growing through the trailer and in the boat and everywhere. So he wasn't doing much work.'

Retired cray fisherman and Torquay identity Ray Milliken remembers Gary as an 'honest, capable man who thrived on an argument. He was well known along the coast and in the fishing industry' and was known as 'Ten Bins' along the coast from Mallacoota to Portland. This was because even if the catch looked bleak, he would confidently predict he would get his ten bins of abalone that day.

Tassie Warn remembers Ten Bins as 'a bloody funny man. How did he get his name? He was at Mallacoota and the boys were catching a lot of fish. He made a statement predicting that someone will get ten bins soon. "Nah, bullshit," we said. And then they started arguing about it. He never said, "I'll get ten bins", but the name just stuck.'

According to Tassie, 'Ten Bins looked after the dive shop in Melbourne when we set up the industry in Tasmania around November 1962. Rick Harris was also there in the shop. I was a mate of Ten Bins for a long time. He was a real character, not just as a diver but because he had some radical ideas. Sometimes I'd go up to his place to have a cuppa. We'd be talking for only two or three minutes before he'd call out to his wife Bev, who could be anywhere. He'd be down the backyard or anywhere and yell, "Where the bloody hell is that cup of tea?" He'd yell as if we'd been waiting for it for an hour. He was a pretty explosive sort of bloke. But a real funny bloke.'

When Frank Zeigler was in the police force in Portland, Gary would sometimes have what Frank called 'full moon nights', when he went a little crazy. None of the other police officers could talk to Gary, because they weren't divers and didn't know what he was on about. Frank says that if it was possible for one person to have three opinions, that was Gary. He could argue about many issues and when Frank would walk over to him in the pub and ask, 'You don't really believe that do you?' he'd say, 'Shit no!' He was a loveable fellow, says Frank: 'You could have a laugh with him.'

Frank recalls that Gary's son Kirk only ever wanted to be a diver like his dad. But Kirk had a range of problems because of a debilitating diabetes condition, and Gary was very protective of him. He was aggressive towards people who attacked any member of his family, especially Kirk.

Frank says rumours circulated that Gary hadn't died in that boating accident and there's been a few sightings, 'But no, the fact that his body was never found is completely normal when you're lost at sea,' Frank says.

'There have been a lot of bodies they haven't been able to find.' Besides, he knows Gary wouldn't have been able to keep away from his sons. 'If he's still out there, then someone would have seen him near Morgan or Kirk. He was too close to his boys to stay away.'

Rick Harris acknowledges Gary as one of the very early divers in the industry. 'The stories he'd tell — that's why they called him Ten Bins. He reckoned he was the best diver, the best everything. He was scared of sharks, he was *really* scared of sharks. And he'd only work when he had no money left. If he had money he wouldn't go to work. When he did work he'd spend all his time sitting in the boat on top of a good patch of abs.'

Rick says Gary was easily distracted. 'He used to go to the wharf and block the boat ramps so the divers couldn't get their boats in the water. He'd park his boat right across the boat ramps. He didn't want to go out, and he didn't want anyone else to go either. He got nicely bent a few times and he'd spend all night under the shower sucking Bex painkillers. He'd eat them like lollies. He'd have a Bex and a cup of tea. Half the night he'd be under the shower throwing down Bex like they were lollies.'

Paul 'Noddy' Hill

'I was a mess after the guys died'

Noddy's writings from the mid 1980s and a video about him made around 2015 reveal him to be an intelligent and deep-thinking yet solitary man. Many of the Portland divers remember him as a reclusive and somewhat eccentric man, but recognise him as being one of the pioneers of the industry in Portland.

It's said Noddy earned his nickname because he was known to nod off to sleep after a few drinks. He was born in Sydney and had a troubled early life, and for 10 years worked as an electrical technician in factories producing domestic tools and machinery before becoming an abalone diver. He saw this lifestyle as a lucrative way to make good money.

Noddy was among the first wave of abalone divers in the Western Zone and recalls diving in those early days with other pioneers such as Rick Harris, Rod Ussher, Tony Jones and Dick Kelly. Blue Grant recalls Noddy as originally being 'one of the Mallacoota divers'. Noddy was mates with Derek Fieguth and he has very fond memories of Ten Bins, Yogi Braun and

Konrad Beissen, who he remembers as a 'good bloke'. He recalled how devastated he was when Bernie Morton suicided.

Noddy's favourite dive site was The Passage, although he says the abalone were not that big there. Unlike many of his diving contemporaries Noddy did not have a big shark encounter, and he didn't like diving Lady Julia Percy Island. 'While it was good diving there, for me it was terrifying. I'd always be looking over my shoulder for sharks.'

He said: 'I make no claims of being any sort of a gun [diver],' maintaining he was mediocre. He recalls earning $1168 in four hours of diving and that he made 'about $60,000 with a little $8000 boat'; on one occasion he brought home 14 bins of abalone. He recalls the days when he would work a section of a reef for seven or eight days straight and how this changed during his time to the point where a reef could only be worked for one or two days. Noddy said that at the time of the consolidation of licences the 1984 licence fee was $2620, with a transfer fee of $5000 plus $10,000 per consolidated licence.

In his many years as a diver Noddy recalls how his catch declined significantly because of over-fishing of the reefs and divers taking 'undies' (under-size fish). He recalls not being willing to dive deep as he believed it would rot his bone structure out with bone necrosis. He said all he wanted was a decent wage from the industry. Noddy was in the abalone industry for 16 years before losing his $90,000 licence for not paying his licence fee, despite an offer of a loan from other divers. He said that by the end of his diving career he was fed up with the industry.

The Western Zone has only had two deaths among its divers, which is surprising considering they dive in notoriously dangerous and shark-infested waters. The only deaths were both deckhands for Noddy and both occurred in the seas off Cape Nelson within five months of each other.

The first death was that of Allan Barr, aged 20, who died in Noddy's arms after they had problems with Noddy's boat. According to the *Portland Observer and Guardian* of 3 August 1970, Noddy 'fought smashing surf and physical exhaustion before being rescued' 19 hours after his boat sank. According to the report, Barr's body was never recovered. The article goes on to say: 'Hill, an experienced diver, on Saturday

lost his new boat and equipment, worth more than $3000. It was insured. His new vessel, a fibreglass runabout, had every measure necessary, plus a few more. Hill had, in the past few days, considered installing a two-way radio in the boat.'

Tragically, five months later Noddy's sheller, Frederick Heilien (known as 'German Freddie'), met a similar fate. The same article in the *Portland Observer and Guardian* tells how Noddy and Heilien became separated

from a group of abalone boats only minutes before a flash storm hit Cape Nelson.

According to the article: 'The motor on Hill's new boat refused to start half a mile [800 m] west of the Cape Nelson lighthouse and about 150 yards [137 m] from the rocks. Five other abalone boats had rounded the Cape to return to Portland, not realising that Hill's boat was washing towards the rocks. Hill and Heilien both donned life jackets, Portland police were told later. Hill, realising their lives were in danger, fired several flares and sounded a special siren he had installed in his boat for safety purposes. A group of people on the cliff top noticed the two men sitting in the boat in life jackets and waving. Police said yesterday it was almost 6 o'clock on Saturday afternoon before they received the message. It was more than an hour earlier when the boat went onto the rocks with Heilien still aboard — he had refused to leave the boat. Hill had bailed out with his life jacket and tried to get Heilien to join him. The last he saw was the boat going onto the rocks. Neither Heilien nor the boat had been sighted late yesterday because of huge seas which made searches impossible ... Hill's motor refused to start and his vessel, himself and Heilien were left to the mercy of the sea ... Hill was brought back to Portland and District Hospital in a state of severe shock. He was extremely upset.'

The article went on to reveal more about the tragic events leading up to the deckhands' deaths: 'Saturday's tragedy was the fourth in which Hill, 29, has been involved since coming to Portland. He suffered an attack of the bends at Cape Nelson two years ago and in March almost drowned when his compressor failed. This happened about two weeks before he lost his sheller Allan Barr in his arms, in a sea cave.'

These two drownings still haunt Noddy and he becomes emotional when asked about them. 'I was a mess after the guys died,' he said. Noddy dived for seven years after the drownings, his final dive being in Portland in 1987.

Noddy's life became complicated after he left the industry as he was involved in disputes with various government authorities. He lost everything and retreated to a reclusive life at what he called Placebo Park, his campsite deep in the Otway Ranges, where he lived for around 15 years. He called it his isolation from the world. There he built a series of eco-friendly huts and only ventured to nearby towns to buy necessities. He set up water and power supplies plus internet access to enable him to maintain contact with the outside world. From the Otways he wrote on such subjects as income and capital gains tax, land speculation, art and antiquities, domestic violence and Noddy's Law. He had no formal education and self-educated himself across many science disciplines. Noddy admits he didn't want to live in society and loved the tranquillity and peace that came with living alone in the forest, but that his lifestyle was incredibly lonely.

These days Noddy lives in an assisted-living facility in Geelong and is only a shadow of the spirited young abalone diver who became a self-imposed recluse. Noddy suffers hearing loss, speaks barely above a whisper and sometimes drifts into his own thoughts mid-sentence. However, chatting with Noddy stimulated memories of his abalone diving days — some good, and some not so good.

On 26 September 2015 the *Geelong Advertiser* said of him: 'To most people who claim to know him, "Noddy" may be many things, but he is not dangerous; he is a strange hermit, a brilliant eccentric, a misunderstood genius …'

You can get an insight into Noddy and his life as a recluse by viewing the short video at https://is.gd/Im38Yk.

Details about Noddy's life and diving career are sourced from his own writings: *The Standoff Story* and *Capital Gains*, and the above video.

7. The divers

Konrad Beinssen

A good lifestyle decision

Konrad Beinssen's introduction to the abalone industry was from an academic perspective. As a marine biologist, he worked as a scientist with Victoria Fisheries for nine years from 1968. After a stint sailing, he returned to the industry and has been involved with fisheries science ever since.

Konrad undertook research in the Western Zone, including two major experiments at Cape Nelson very early on. One of these was to measure catchability constants, which is a supposed constant between effort and fishing mortality. The premise is that if you know this constant, you can measure the fishing effort (ie, the number of hours dived), then the percentage of abalone that is taken from the population equals the fishing mortality. Establishment of this constant is important for use in mathematical modelling of the fisheries.

The second project was known as the 'coke can' experiment. Thousands of normal-size drink cans without tops were filled with concrete. A loop was placed on each and three tags were attached. The cans were distributed along the Western Zone reefs. Each time they passed over one, divers were asked to take off a tag and return it to the researchers. From this, it was possible to estimate the reef coverage of the abalone fleet, that is if the catchability of cans equals the catchability of abalone. That way it is possible to determine the fishing mortality. This experiment was conducted in the early 1970s and the results have been published in scientific literature. Although not all divers

Clockwise from top left: Abalone catch, abalone sizing. Phil Sawyer and Phil Cuneen shelling at the Safcol Wharf, Portland. Phil Sawyer driving the hydraulic bag carrier, which could support two divers and could fit 400 kg of abalone at a time. *Photo source*: Konrad Beinssen

assisted in the research, the results provided a good estimate of reef coverage by abalone divers.

Konrad says divers in the Western Zone were particularly good at assisting the researchers and that many meetings were held to explain the importance of the research to the divers. The results of these experiments were for general modelling of the fishery. A lot of work on growth rates was also conducted and many abalone were tagged from various reefs to measure growth rates.

From 1989 to 2004 Konrad dived for Phil Sawyer, who had two licences. Although there was a lot of money to be made, it was the freedom offered by the lifestyle that attracted him to diving. 'Once I'd caught the quota, then I could do other things. It was a good lifestyle decision.'

He remembers the divers causing social problems in the small fishing communities. Those outside the industry resented these 'reprobates earning a hell of a lot of money and only working a few days of the year'.

While working with Phil, Konrad was able to continue his research. He saw Phil as an innovator and recalls how he used to have a hose over the side of the boat and the boat's motor pumping water through a gas heater and the warm water pumped into Phil's wetsuit. 'One day the hose floated past a group of jellyfish and it sucked some into the hose. It mashed them up and pumped them into his suit. He made a very rapid exit!'

Konrad describes the diving cage Phil bought from South Australia, where abalone divers commonly use them. Based on a design by a man who worked out of Port Lincoln, the diver remains in the cage and collects abalone through the bars. Phil adapted this cage by taking off the bottom and modifying the bag carrier. This adaptation meant there was less need to keep returning to the surface with the catch. A normal diver's bag weighs perhaps 80 to 100 kg, whereas the one on the cage could hold around 400 kg. Another advantage was the cage was driven by a propeller, which meant it didn't need to be dragged along. 'Phil and I worked the one apparatus. We'd go along and park it on the bottom, then do a sweep round our umbilical cord and pick up what we could. We'd put these into a small bag and then return to the apparatus and put them into the bigger bag on the cage. It was much more efficient.'

In the early days of the industry, hookah diving was used in Eden/Mallacoota and divers started off by using garden hoses and 'dicey' compressors. Before that they used scuba but it was inefficient because they had to get their tanks refilled and carry bottles on board their boats. 'I ended up using nitrox, a nitrogen/oxygen mixture, for my last years of diving.'

Konrad explains that if you breathe oxygen from below one atmosphere, it's possible to have a convulsion. Oxygen becomes toxic at some point but some divers have a better tolerance than others. You can't breathe pure oxygen — if you could, you'd have no problems with the bends. Nitrogen forms the bubbles in tissues, so if you enrich the air with more oxygen, normal air has 20 per cent oxygen, to make a 40 per cent mix of oxygen and nitrogen, then you can lower the risk or extend the time you spend underwater.

Being a researcher, Konrad knew the value of submitting accurate catch returns and says he was more diligent in filling these out than some divers. 'I understood the dynamics and I ended up producing a bio-economic model of how the industry worked and this was published in one of the Fisheries journals. I was always aware of the dynamics of how the industry could be sustained and the problems associated with overfishing.'

Konrad believes there is a future for the abalone industry. 'It's an interesting industry that's still near and dear to my heart, even now. It's going through tough times but it will come out as long as it is carefully managed. Abalone fisheries around the world have a history of collapse. Hopefully this one will return to productivity.'

Konrad now lives in Queensland.

Dennis Carmody
Cruising with a Land Rover, a girlfriend and a fast boat on the back

Dennis Carmody was in the second wave of divers who arrived at Port Fairy in the mid 1960s and he stayed in the industry until 1980. In the early days sports diving was just gaining popularity. Dennis had been learning to dive at the Melbourne City Baths with a couple of friends. 'The industry was in its infancy and we'd heard rumours about the money to be made. All we'd done at that point was scuba up and down the length of the baths, but we thought we were going to be rich.' He and Harry Bishop bought a small boat and headed to Phillip Island.

When they got there, they met a Russian chap called Les. The story goes that Les got into the industry after answering a classified advertisement for what he thought was a 'driver' but it turned out to be for a commercial 'diver'.

Before Dennis and Harry could start, they knew they needed to learn about abalone diving. Les took them to Lookout Point at Seal Rocks. 'Well, there's my boat out there, boys,' Les said. 'Swim out and introduce yourself to the team. And we did.' That was their induction to abalone and the industry.

They dived at Phillip Island for a short while and then made their way to Port Fairy, where Dennis stayed for some time. Life was sweet in those halcyon days and Dennis reckons every day was an adventure. There was little specialised equipment and what they did use was 'dodgy' to say the least. They used old fridge compressors and Dennis recalls how their wetsuits were usually falling apart.

The strong seas and bitter winds arrive at the Western Zone straight from Antarctica. 'Some days it was so cold that you'd have to put your wetsuit on in the morning under a hot shower in the camping ground. Once you were in, you were in for the day.' The waters are also notoriously treacherous, with perils below and above the surface. According to Dennis, 'Once you hit Gabo Island and go west, that's real diving. It's colder. It's stronger. It's bigger. Even the crayfish come out of holes and try to get you.'

On their first week diving, Dennis and Harry were working in strong water when Dennis saw blood all around them. He recalls thinking at the time, 'There goes Harry' — they had already been spooked by their lack of diving experience and the large number of seals around. Dennis surfaced and jumped straight up and into the boat at the same time as Harry. He reckons he could have got bent because of the speed he came up. 'Then we looked around and there was this film-maker tipping 44-gallon drums of ox blood into the water to attract sharks around us for his film. We followed him in and had a "strong altercation" with him on the boat ramp.'

Dennis acknowledges all divers have their shark stories, but he now sees these stories as boring. 'I don't think the sharks want to know you.' An undated newspaper story titled 'The Mystery Disease Divers Fear More Than Sharks' has a chilling quote about sharks in its introduction: 'The abalone divers who descend into the silent, green depths near Julia Percy Island off Portland, have a saying: "You'll never see the shark that gets you".' Several abalone divers have been taken in recent years. The most recent was in February 2011 when two great white sharks attacked and killed an abalone diver at Coffin Bay in South Australia. Dennis maintains that this is an extremely small number of fatalities considering the amount of time abalone divers spend in seas that are notorious for sharks. 'If you're looking for sharks, then you're not working and shouldn't be there.'

According to Dennis, there was always good camaraderie mixed with a bit of competition between the Port Fairy and Portland divers, but nothing ever turned unpleasant. He recalls how Len McCall was a legend for knowing all the words to 'Drop Kick Me Jesus Through the Goal Posts of Life', a badge of honour in anyone's language.

'These were such good times that I reckon I've been paying for it ever since — a sort of yin-yang thing. It was a lovely period of my life — all wine, women and song. I had a Land Rover, a good girlfriend and a fast boat on the back. We just cruised.'

Like many of those in the first waves of divers, Dennis missed out on the big money that was made once they rationalised the industry and he ended up losing his licences in Victoria and New South Wales.

Dennis admits he's a bit of a butcher when it comes to preparing and cooking abalone. First he cuts everything off that looks nasty, such as the 'pooper' and the outside edge. Then he slices it across the flesh horizontally. He uses a cook's large tenderising hammer and gives it one bash on either side. He then dips it in garlic, egg and breadcrumbs and cooks it for 30 seconds. He says, 'People love you when you do it this way.' He has also had it cooked professionally in the shell and admits that although that's nice, he prefers 'my messy wasteful way'.

Jason Ciavola
The solitude of diving
Jason Ciavola is one of the younger of the new breed of divers working the Western Zone. He is a commercially qualified diver and works multiple licences.

Jason was born in Melbourne, but his mother is from Port Fairy and they moved back there when he was ten. As a young adult, he spent five years in Western Australia working and surfing. He also worked in Queensland harvesting the highly prized sea cucumbers known as *bêche-de-mer*: 'They taste awful but the Asians love them because of the texture,' he says. He likes working in the abalone industry because of the independence and autonomy it offers. 'You don't have anyone looking over your shoulder.'

When he returned to Port Fairy Jason worked as a deckhand until he started harvesting abalone. After leaving and doing some commercial diving, Jason worked for David Forbes as a deckhand and as a diver until he starting working Len McCall's licence in 2005, the year the virus came. 'It was a disheartening year. It started off terrific. Then the virus came within a month of my starting and that changed the whole dynamics.'

They lost about 8 tonnes of quota that year with the quota cuts. The following year he worked a very small quota and picked up dribs and drabs of extra work while continuing to work for David Forbes, 'I do a little bit of research work for David Forbes from time to time and go to meetings down at the labs in Queenscliff.'

Jason enjoys the solitude of diving. He says that divers work shorter days in the deeper areas, as they can't spend as much time below because of decompression illness. In places like Julia Bank, he says, they can only spend about three to four hours in the water, whereas at somewhere like the Island, they can dive shallow and stay six to eight hours. 'I usually work about a six- to seven-hour day. That's enough for me.'

Jason does both deep and shallow diving. 'I do pure oxygen decompressions as well. I decompress with medical-grade pure oxygen at around 9 or 10 metres, and that flushes the nitrogen out of my system. I come up very, very slowly. The slower the better,' he says, 'then I usually try and find some shallow ground because I don't like hanging off the boat, especially at Julia Percy Island or the Bank because there are big things that swim around there. But I will if I have to.'

Three hours is probably the most time he spends underwater without a break. 'By then I'm usually hanging out for a drink and a smoke and a chat to someone.' He doesn't have a big meal, just a nibble.

Jason uses a drop-line system where a weighted bag on a line gets thrown down to him while he is still working. He removes the empty bag, clips the full bag on, inflates the parachute and it goes to the surface and is pulled into the boat by the deckhand. This means he doesn't keep going down and then resurfacing — or 'bouncing', as he calls it. But he gets the occasional seal that wants his bag.

'You try to warn him off but if he's a cheeky one, he wants to beat you to the bag and will try to swim off with it. He'll snap at you and have a bit of a play. Divers always say they see sharks,' Jason says. 'I'm sure they see us, but we see very few of them. Some divers have been diving for a longer time than me, recreationally and professionally, and they've never seen one. But there are sharks out there.'

There are lots of colourful characters in the abalone industry, according to Jason. At meetings, he recalls hearing 'clink, clink, clink', because one of the divers would be smuggling beers in. And there was often the smell of a joint wafting in through the door. On the same subject, there's a much-repeated story of how one diver used to get his deckhand to blow the smoke from joints down the hose line to him.

Jason would like to spend a few more years in the industry. Ideally, he would like to sell his other business and just dive for abalone. 'The last few years have been too hectic. The kids are growing up and I hardly see them because we're always waiting for the sea. I'm under a lot of pressure to get quota caught. I haven't got just one bloke I'm working for, I've got three or four to keep happy. It's all right if they're ex-divers because they understand. If they're not, they get a bit grumpy.'

Jason likes eating abalone. He slices it thinly and coats it in seasoned flour. He knocks as much of that off as he can and shallow fries it really quickly. He usually has a mixture of garlic and chili, coriander and basil, which he sprinkles on top, then folds the abalone in half and eats it. He doesn't believe in tenderising it first: 'The idea is to eat it as it comes out of the pan. If you let it cool down, it will go tough.'

Robert Coffey
Good divers, good businessmen — or both

One of the second generation of divers, Robert doesn't dive commercially any more, but still dives recreationally. He dived on his father Andrew's licence between 1992 to 2002.

Robert didn't think much about growing up in a fishing family, but, 'If he hadn't been diving for abalone, Dad would have been diving anyway. He was an incredibly keen recreational diver and a very good spear fisherman.' As a child, Robert recalls he always felt safe when he was diving with his dad. 'In my very early diving days, I'd be catching crayfish and dad would be catching abalone just off the shore. We had an inner tube with a potato bag tied to it and he would throw the abs in there.'

Diving with his father was a great apprenticeship and he learned a lot about the sea, including the importance of knowing how to swim in a seaweed-infested area. 'You can move through the weed without entangling yourself,' he explains. 'But when you're in a rush to get to the surface because you're free diving, you want to be able to clear your snorkel of the water and take a breath.

'However, on the surface, the canopy can be anything up to 2 feet [60 cm] thick, so you have to come up with your hand attached to the top of your snorkel and you have to be able to poke it through the canopy to clear it in order to get your breath of air. If you struggle it's going to rip your mask off and you're still going to be 2 feet short of the surface. So it's important to break the surface with your snorkel, take a breath, clear it and get back under the water and clear the thick canopy and get out of there. An inexperienced person could drown in a situation like that.'

Robert was a disillusioned secondary school teacher around the time his father said he was going to sell his licence. Robert talked it over with his wife and they decided they would buy it from him.

But first he needed to learn the ropes. Up until then, he had been diving off the shore and he became Andrew's deckhand to learn about compressed air and boats. Robert enjoyed the industry while he was involved in it. He particularly enjoyed the freedom. 'In a quota-managed fishery, it wasn't so tough if you missed a day's diving because you could catch up. But if you don't have a quota-managed fishery then a day lost is a day never regained.'

For Robert, the abalone industry offered life-changing opportunities. He was a representative for WADA, secretary for three or four years and president for about six years. He says it was an exciting time politically, in terms of how the industry was trying to position itself and because of the politics between the Zones.

'We were part of the next generation of divers. It was a change-over time, it really was in a state of flux. When I first got in the water, my father was still diving. Murray Thiele and Jurgen Braun were both still diving. Nearly all the Portland divers were still diving. Probably Dick Cullenward was the first one to go. Then there was a progression: Len McCall got out and Jurgen Braun, so the only original people still diving were Murray Thiele and Rod Crowther. Rick Harris was only there for another four years before he got out.'

In Robert's opinion, abalone divers tend to be strong individuals, particularly the early divers. The people who began, and those who survived the early days and continued on, he says, were a combination of good divers or good businessmen, and sometimes both. The ones who weren't didn't survive. Some of those who were cowboys survived and some didn't. Robert explains, 'They weren't concerned with the risks involved. They had no idea what the long-term effects would be and in the end it seems as though they were pretty much vindicated. A lot of them continued diving well into what most people would consider their senior years. That sort of cardiovascular exercise was clearly beneficial in many ways.'

Robert claims he didn't see too many sharks and doesn't believe anyone who said

they saw a lot. He did most of his diving at Lady Julia Percy Island and that's where he ran into sharks, although he also saw them at The Crags.

His first close encounter was a murky day and he was down about 18 metres. He was chipping at abalone and putting them in the bag when he looked up and all he could see was what he describes as a 'black-and-white ABC test pattern' on the shark's skin. 'It was massive, like a 44-gallon drum, and there's no way known you could get your arms around it. It was right there but I had no idea where its head was.' He knew it had a good look at him. 'It was eyeballing me and wondering what the hell I was.' The shark steamed past him and turned around and came back to look at him again. Robert says he turned and looked at the shark and, with that, it flicked its tail and was gone. It was over almost too quickly for him to panic.

The second time was different because he felt he was being hunted. He was at the Island, in his territory and working his way back to the boat, and there were two other divers close by. He remembers crossing some sand and all he could see was a massive tail that would have been a metre and a half. He had time to think, 'What do I do here?' He swam to a large rock and looked at the shark as it was going away, then went to the other side of the rock thinking that if he couldn't see it, then perhaps it couldn't see him. 'But, sure enough, the damn thing came around the rock, which was about the size of a room, and was coming back towards me.'

He found a crack in the rock and went straight into it, like a seal. 'The shark came by and eyeballed me. I counted to about 60 and hoped it had gone. I came out and there it was — straight above me. It was a big shark, a really big shark, just waiting, because it knew that seals eventually run out of air and have to get to the surface. I backed into the crevice and I counted again and the damn thing comes round again. I counted another 60 and emerged. No shark. I did a lap of the big rock and I saw it swimming away. That was scary.'

Robert stopped diving because, from his reading of what was happening in the industry, it was a good time to get out. 'My father had been indicating for some time that he'd be happy for me to consider selling the licence. There were a lot of things happening politically as well as internally within the industry. Forces were massed against us, so there were a couple of compelling reasons to sell.'

Rob's favourite way to cook abalone is to fry it in breadcrumbs. He cuts it thinly horizontally and lightly pounds it, dips it in egg and breadcrumbs, then into the fry pan. 'It's a rich flavour,' he says. 'There are other ways of doing it. You can treat it like a steak. You tenderise it and then cook it like a blue steak — fast and furious on a very hot grill, and that's it. But it's like everything, if it's common, it's not highly regarded. Like the difference between a weed and a flower, it depends on how much you like it.'

Jamie Espie
It's competitive out there

Abalone diving is in Jamie Espie's blood. His father Clarke was one of the early divers and he encouraged Jamie to give it a try. Jamie was 24 at the time and probably one of the youngest divers around. When he heard a whisper that Rod Crowther wanted to 'get out of the water', he recognised that 'the opportunity was there and I gave it a go. It was like a holiday going down there to Portland. It was a lifestyle thing then, but it has changed a lot. These days you have got to have another job to sustain you.'

Jamie dived the Western Zone from 1994 to 2004, diving for Rod Crowther for about five years and then for Rick Harris to catch part of his quota. 'I was one of the youngest, then a few of the older divers decided to move on. I dived differently to most because quite a few used to anchor up. I did a little bit of this when I first started, but once my deckhand got used to things, we stopped using an anchor altogether.'

Jamie says Rick was terrific to work for. He'd stay at Rick's small house at Cape Bridgewater and 'quite often he'd be home cooking some sort of feast and he'd ask us around to eat'. Jamie remembers 'Rick used to wear Giant brand flippers and he'd cut out the fin so it was like a shoe. They are like the Croc shoes they wear now and looked like they'd been bitten off by a shark. It was a pretty good idea — he should have marketed them.'

The wild waters off the Shipwreck Coast are notoriously unpredictable, with fierce conditions prevailing for much of the year. There's not a lot of coast in the Zone compared with the other two Victorian zones but, according to Jamie, it still takes years to get to know the coastline and patches. Jamie was aware that he was competing against wily, experienced divers who didn't readily give up their hard-earned knowledge of the seabeds. 'If you are new like I was, you don't get time to swim the whole coast in deep water, so you keep an eye on the older divers to see what catches they have. The deckies learn this knowledge, which is helpful.'

According to Garry Bartle, Jamie was an 'awesome' diver. 'One day I said to him, "Look, mate. Give me a few bucks extra and I'll show you where all the abs are." He just turned to me and said, "I'll find them".' Garry worked ad hoc for Jamie for a couple of years and reckons he was 'just dynamite',

one of the fastest divers he'd seen, and he used to catch his quota sometimes in four to five months.

Jamie is now in his 40s and dives in the Central Zone and admits that while he enjoyed diving at Portland for 10 years, he never thought of staying there because the Mornington Peninsula is his home. But he learned a lot from watching the ways of the experienced Western Zone divers who operated during those ten years. 'It's competitive out there when everyone is trying to get a big catch for the day. Phil Sawyer and Bob Ussher knew the good spots in the deep water. Rick Harris would often launch after us but he'd always be pulling his boat out even before we came in, often with a pretty good load. No one knew where he'd been. He was quite sneaky about it. I always kept an eye out for where guys like him had been. After a while you become crafty because you want to look after certain areas. Some patches you can always get a good catch, but sometimes it may only be once a year, so you've got to make sure no one else fishes it.'

Accurate catch rates are always tightly held among the divers and Jamie says divers wouldn't tell other divers where they'd been getting their catches. 'Everyone had their favourite spots but a few of the divers didn't know about my favourite spots and I wanted to keep it that way. Sometimes I wouldn't go to my favourite spot if there were other boats there. I wanted to keep it to myself.'

His biggest catches? Jamie reckons in those days he pulled 1400 to 1450 kg. His favourite spots were around Cape Bridgewater while Portland was his home ground. He also liked The Crags and says he would get 1000 kg every time he dived there: 'If I didn't get that, there was something wrong.'

The 'good old days' were still around when Jamie arrived in the Western Zone and the lifestyle some of the divers pursued came as a shock. 'When I first got to Portland, I wasn't too sure what I'd got myself into. One day I was in Rod's boat and I saw a boat chugging along with two of the well-known Portland divers on board. I could see a waft of smoke coming from the cabin where they were. The deckhand was sitting out the back, but the guys were getting pretty stoned before they even got into the water. We went over and had a bit of a chat, but I didn't know what to make of it all.'

Although Jamie admits there were drugs around, he believes that to last as a diver you've got to look after yourself and keep fit. 'Smoking and drinking knocks your body around too much.'

He admits that the lifestyle is the big attraction of abalone diving. He's always enjoyed being in the water and likes having the freedom to choose where he works. However, he says it's a strange sort of life and you have got to be a hard worker to be successful. 'It's pretty hard to describe. When you're a diver you've only got a deckhand, so there's not a lot else to worry about. It's a simple line of business to run.'

In the Western Zone, nearly all of the 14 access licences are dived by someone other than the licence holder. This means the work practices of divers must be professional, responsible and accountable, and they can't afford to make mistakes. As Jamie maintains, 'You can't afford to fool around too much. Licence owners want someone they can trust and so you're always trying to keep the boss happy.'

Jamie does eat abalone, but not often because divers aren't allowed to have any left in their boat at the end of the day. He slices it thinly, adds soy sauce, butter and garlic then throws it on the barbeque.

David Forbes

You've got to have the mongrel in you to be a diver

David Forbes enjoys recounting his worst abalone fishing experience, a story that's proof of just how treacherous and unpredictable the wild waters of the Western Zone can be.

It was 8 June 2004. The sea and conditions at Cape Bridgewater were perfect. There was no wind and the seas were dead flat. David reckons he's never known it flatter. Best of all, he and his deckhand Glenn Harman were working a patch of abalone the likes of which he'd never seen before. There was a tonne of abalone laid out before him. 'If I could have used a shovel I would have.' So rich were the pickings that when David surfaced for lunch, Glenn had five to six bins behind him that were waiting to be cleaned and stacked.

The men had a quick lunch and then got to work to stack the backlog because David knew how much abalone was waiting on the bottom. He recalls it was so flat that the bomboras were 'just boiling out of the water'. He was filling bags with approximately 40 kg in only four minutes. He was underwater when he heard Glenn rev the motors and heard a rumbling sound coming towards him. Everything went white. He remembers thinking that a big wave must have come out of nowhere and the next thing he thought was, 'Gee, Glenn's gone.' Not long after, David heard the motors revving again and realised Glenn was okay and that he was working to get them back to safety.

'I grabbed my gear because I knew something was wrong. My air hose was towing me along the bottom because Glenn was running the boat and trying to get over the wave. Then I heard a giant rumble and suddenly it was like a washing machine. I was only at 15 feet [4.5 m] but it was tumbling and churning. All of a sudden the hose went from going out to sea to heading in towards the rocks and I was getting pulled along. I ended up on the surface. All I could see was the boat sideways on the wave with Glenn standing up to his waist in water. Bins and stuff were floating everywhere' Everything got carried in and slammed against the cliffs. Apparently there were two waves. The second picked the boat up and dumped it into a rockpool. David's boat, *Absession*, was sitting in the pool against a cliff and there was no way to get it out.

If they had a helicopter, David thought, they could have lifted *Absession* up and put it back into the water, and the boat would have been undamaged and fit for work.

But they didn't have a helicopter and the men had to leave *Absession* at the rockpool. The next morning David organised a crayfish boat to pull the boat from the pool. David had to swim in, secure the boat and then wait for the tide to come in. They timed it perfectly, because *Absession* just popped out of the pool. It was pock-marked with thousands of small holes along the keel and had a large hole in the bow. Amazingly, most of their bumper abalone catch was still on board.

The cray boat towed *Absession* back to the shore and the men and the boat were left on a remote beach under a cliff. They grabbed all the important gear, including the Fisheries notebooks and oxygen resuscitation gear, and started climbing up the cliff. 'By the time we got to the top, all we had was the boat hook, and that was only because it was a good walking stick. Everything else got left behind. We had to walk across the cliff, hitchhike home and that was the end of *Absession.*' David's pride and joy these days is *Absession 2*.

The first time David went diving with his uncle at Barwon Heads, 'He took me out with a snorkel, plastic bread bag and a screwdriver, and in an hour I'd caught two abs. I swam back with the ab in the bag, but the bag ripped on the way in and I lost my catch.

'From memory, they used to get £1 [$2] per lb [0.45 kg] for blacklip and £1/3/0 [$2/60] for greenlip. One day they were selling their catch to the fish processor when a Fisheries guy came up and told them they couldn't fish for abalone any more without a licence, and

Clockwise from top: Andrew Beauglehole and David Forbes. Recovery of David Forbes' *Absession*. David Forbes' *Absession* beached. *Photo source*: David Forbes

that a licence would cost them £2. He was told to piss off and they didn't buy a licence.'

The lure of abalone was always there and David always knew he would work in the industry. He decided to volunteer at the Marine Science Laboratories at Queenscliff and got a job as a deckhand before moving into abalone research and husbandry. 'This allowed me to move inside the inner sanctum of the abalone world. I got to see what the fishermen did, how much they earned and the lifestyle they led. I knew it was for me.'

He tried to buy into the industry around 1997 and even had a handshake agreement to buy a licence in the Eastern Zone for $2.8 million. He still shakes his head at the deal he nearly had, which included the licence with 20 units, a boat, a caravan on a block of land at Mallacoota, a tow vehicle and Co-op shares. 'The guy was just walking away. Five years later the licence would have been worth $6 million, the Co-op shares would have tripled in value and the land would have been worth heaps. While the bank would have lent me the money, the bloke simply changed his mind.'

He then got a handshake agreement on a licence in the Western Zone, which was more expensive and without the extras. This was around the time the prices dropped from $45 to $32. But this time the bank got cold feet. David eventually started working for Scott McRae in June 1999, diving his quota for a few years before switching to dive for Bob Ussher.

David has researched diving practices and worked out the best way to maximise his time underwater. 'The mentality when I first started was to do your first day's diving deep, say 100 feet [30 m]. The next day you'd only do 80 feet and the next day 60 feet. Modern thinking is the reverse — start shallow and prepare your body by working up to going deep. They say that with more exposure you are less likely to get the bends.'

A 'pony' tank is a small tank divers carry that allows them to safely surface in an emergency. Unlike many of the divers, David wears his all the time. He says the other divers make comments such as, 'I only wear a pony at 60 feet,' but he thinks 'that's crazy. Can they make it up safely from 54 feet and decompress if they need to by holding their breath?'

Teamwork and a good deckhand are integral to David's operations. 'People say to me that I'm a good diver, but I'm just part of a team. The best diving and the best catches I've ever got are when my deckie and I are working well together.' For a while David had two deckhands, Phil Haywood and Andrew Beauglehole. 'One would drive the boat and the other would pack. We'd be pulling a tonne most days so it was busy on deck and they'd just keep the boat moving and the bag changes going every four minutes or so. It worked like clockwork. We were a team. One day we got 1480 kg at Cape Nelson.'

There's a well-known side effect of diving that David calls 'diver dumbness', which is a 'doughie' feeling in his head as if his head is filled with custard. He doesn't know whether it's because his body is still full of nitrogen and doesn't have enough oxygen after a dive. He doesn't know it's setting in when he's diving and finds that, at the end of the day, he feels thoughts coming in one side of his head and eventually out the other. 'You quite often sit there and roll your eyes and think, "What was I just thinking about?" You make simple mistakes on paperwork, like when adding up catches on the day.'

As well as Phil Haywood, who worked as David's deckhand for five seasons, and Andrew Beauglehole, his other deckhands included Glen Harman, John Russell, John Warne and Chris Sutton. The other divers would joke, 'Burned out another one have you, Forbsie?' Phil Haywood never wanted to be a diver and sums up the attributes required of a diver perfectly when he says, 'You've got to have the mongrel in you to be a diver and I just don't have it.'

Phil has a lot of respect for David, both as a boss and mate. He describes him as a determined sort of bloke, who's particularly hard on himself. Even when the weather was 'ordinary', David would still go out and try to catch something to justify his drive from Geelong to Portland. 'He's got a very strong work ethic. He's pretty focused.'

While other divers use 60- to 100-kg bags plus a parachute, David prefers to use a smaller bag. As he says, you can't take a 100-kg bag into the weed because it becomes tangled. The way most divers work is that they swim to a clearing and work it, then swim to the next clearing and work that. They work the weeded patches by starting off in a weedy patch when the bag has little weight. Some of the older divers worked what they called 'hungry bags', where they'd leave the bag in the clearing and take a smaller bag into the weed, then come out and transfer the abalone to the larger bag.

David uses 30-kg bags, known as 'handbags' to the other divers, which gives him greater freedom to move wherever he wants and fish weeded patches. He also thinks this system is good for the deckhands because they can lift and handle these bags more easily. David doesn't carry a parachute, preferring to have his on the drop line. The deckhand comes along every few minutes and checks what the bag looks like. If it's overflowing, he'll make the change quicker. If not, he'll slow down. He will then bring the boat over the top of David and throw in the line with a stainless steel chain. 'It's amazing because you hear it hit the surface. *Chingggg*. If he's doing his job properly it lands next to me. I take it off, attach the parachute and unfurl it. I fill it and keep fishing.'

David, who quit working the Western Zone at the end of the 2010 season, says the biggest year he ever had in the Zone was 48 tonne.

Craig Fox
Newest kid on the block

At the time of writing this book Craig Fox was the Western Zone's latest licence owner. He purchased half of Phil Sawyer's licence in late 2012 and did his first harvest on his new licence in August 2013. While he is the new kid on the block, his family's involvement goes back to some of the early pioneering days of the abalone industry in Port Fairy.

In the late 1960s Craig's parents, Keith and Kaye Fox, operated a transport business based in Port Fairy, and one of their services

PORT FAIRY — SELECTED FISHING STATISTICS,

YEAR	ROCK LOBSTER (Kilos) YEAR ENDING 30 JUNE	ABALONE (Flesh Weight — Kilos) YEAR ENDING 31 DECEMBE
1965	n.c.	2,026
1966	31,605	n.c.
1967	42,569	17,361
1968	59,108	47,118
1969	58,950	11,290
1970	73,166	6,803
1971	64,579	13,822
1972	70,695	12,734
1973	74,726	6,868
1974	59,519	19,203
1975	33,900	14,415
1976	26,006	n.c.

was to transport the local fish to the Footscray markets. Included in those deliveries was the abalone caught by original divers such as Len McCall, Dick Cullenward and Johnny O'Meara. Kaye still recalls they were paid two cents per kilogram to take the abalone to market.

Craig has been involved in the fishing industry since 1992 and the abalone industry since 1996, after leaving school at age 16 to chase his dream of becoming a professional fisherman. Craig started as a crayfish deckhand with his cousin, David Lane, and moved to crayfishing with local fishermen Digby Wolff and Paul Armstrong. He also worked as a deckhand with David on Jurgen Braun's abalone boat and continued to deck on Jurgen's licence for the next five years.

This work fuelled his passion to become a diver, and in 2001 an opportunity arose to lease Ron O'Brien's licence. Craig took up the licence in 2002 and gave away deckhanding to become a lease diver. 'At the time it was out of my league to purchase the licence as this was in peak price for both licences and high beach prices. Along with 20 tonnes of quota per licence, my dream job and opportunity seemed lost,' Craig said.

Craig left the abalone industry for a number of years after he purchased his parents' transport company, which he still operates from Port Fairy. 'Gone are the days of transporting locally caught fish, as over the years we have seen the decline of most fisheries. We transport mostly grain and fertiliser nowadays, which is a contrast to the early days of Fox Transport.'

Craig always kept a close eye on and interest in the abalone industry, and often relieved as deckhand for people such as Morris Dalton and Cleave Thiele. 'I clearly remember the impact and devastation when the virus hit the zone and I am still shocked at the damage and loss to the local ab industry. Having close friend Peter Riddle still involved over the years was instrumental to me wanting to buy in when the opportunity came along. Peter's advice and guidance reinforced my decision to buy the licence and get back to my dream job.

'It's been seven years now and we are seeing the Western Zone go from rock bottom to a steady incline, and this is testament to the hard decisions made by all in the Zone. We were very fortunate to have Harry Peeters advocating and representing

WADA during that time. What I hope to see, and strive to do, is for the Western Zone return to near what it was when I first started in the industry. I am confident that's happening.'

Craig has many fond and funny memories of the characters in the industry and the great times had by all. 'We used to smile every day when Peter Ronald and Lance Wallace worked out of Dick's sharkcat, *The Double Uggly*. We thought how appropriate the name was and we used to take the piss out of them. It was good banter. Murray Thiele was always reminding me about how Dad wouldn't cart his abalone to market because another diver wouldn't allow it. I guess the theory of the other divers was if Murray can't send them, well, there's more for me.' Craig also recalls Murray telling stories of working Lighthouse Bay, referred to as the Bank, where you would go on a rough day if you needed to fund a new car or the like. 'The abalone were stacked three high on top of each other.'

Above all, Craig knows those in the Western Zone are a great group of people who look out for each other, and he enjoys the diverse challenges of the industry. 'I like working on my own and I enjoy the variables. It's not like work for me. I just love everything about the job, from driving the boat to work to thinking about new places to dive. What I like most is that no two days are ever the same. I like the diversity within the zone and the differences between Port Fairy, Portland and Warrnambool. We have some of the best habitat and reefs along the Victorian coast.'

Since Craig's return he has taken on the role of WADA chairman and is also a director of the Abalone Council Australia. 'I am passionate about the industry and want to be part of fully rebuilding this great zone, and I hope one day one of my daughters might take the diving over from me.'

Graham 'Blue' Grant
Old Piss 'n' Moan

Blue Grant got into abalone while working in the tuna industry. Tuna fishing was seasonal and to him abalone looked like a good thing to do in winter.

He worked out of Eden and Tathra and got as far as Port Stephens. He acknowledges how little he knew about diving when he first started: 'There I was sitting up forward on the gunwale when a big size 11 boot tipped me over and a voice yelled out, "Equalise going down and breathe out coming up".'

He met up with Mallacoota divers 'Noddy' Hill, Bernie Morton and Phil Eather, who came to Tathra to dive for abalone. They shucked at sea and often worked on their own without a deckhand.

When Blue started he thought three bins of meat was pretty good, even though the 'guns' were coming in with eight to ten bins. His catch stayed much the same but he became a better diver just as the resource began to be depleted in New South Wales. The divers had the view that they would not have any management controls and that the resource could never be fished out. Blue was one of the few who believed in a sustainable fishing industry. He had been in the industry off and on for many years and adopted the attitude of leaving little impact upon the environment wherever he worked. 'I do not touch any fish that don't make the grade. I see no point in sending "unders" up to my deckie. I see no point in chipping off "unders", full stop. I prefer other divers to follow me and for them to laugh at the amount of fish I leave behind.'

Blue remembered Ron O'Brien's advice when he entered the dive sector: 'leave fish behind'. Blue believed stewardship was their responsibility and the industry's obligation. 'That's why we have our own diving code of practice independent of government, independent of the fisheries. It is our form of self-policing and promoting sustainable work practices.' He also believed research was vital. He wanted to start a campaign to let divers back into areas they had been banned from, explaining that multiple uses was what they should be demanding.

Blue is remembered for having a unique dive plan. Often when he was out, he'd phone Frank Zeigler, the local dive shop operator. The conversation went along the lines of: 'Well, I've been at 60 feet [18 m] for four hours and done this and that and I'm diving on this mixture and how much longer can I go? What depth am I allowed to stay at?' That was his dive plan — to call Frank up to see what he could do for the afternoon.

David Forbes remembers Blue Grant as a true character and that he had many quaint names for others. David recalls Blue used to call him 'Scott's slave', because when Scott McRae said something, he took notice. Blue used to call Daniel Ussher 'Son of Bob' — that is, SOB. David's deckie Glen Harman made up a number of names for Blue, including 'Back in 'Nam' because he always used to speak as if he were a returned army colonel talking about the last battle.

David acknowledges Blue was a great fellow but the biggest procrastinator in the world. Blue would go to what was known as Umm and Ahh Point, at Portland, where all the divers went to check the swell. Then Blue would go home and check his computer before going back to Umm and Ahh Point. He'd then ring his deckie and say he wasn't sure about whether to go out and would ask what he thought. They'd have a coffee and go back to the computer. Eventually it would be 10.30 am and by then Blue would start whingeing because it was too late.

The other name Blue had was 'Old Piss 'n' Moan', because as well as procrastinating he did a lot of complaining about the conditions. David reckons that name suited him down to the ground. Apparently Blue would turn up, have a look at the water and then start moaning, 'It's a bit big here, bloody this and bloody that.' He'd carry on and eventually get in the water, but he soon got out for a break. He'd do a bit more and get out again, saying, 'What are we doing here? It's too bloody big. We shouldn't be here today.' Then he'd get back in the water and dive a bit more. Others used to go over to him to say hello at lunchtime and he'd be moaning some more: 'Bloody this and bloody that.' Although he'd given up smoking, he would regularly have a cigar for lunch.

Blue used to work big bags and anchor rather than working live. He used nitrox when he dived and then worked for four, maybe four and a half hours. 'That's my day. I surface once. After about three hours I come up for lunch.' Later on he got a newer boat and a deckie who knew how to work live and started using smaller bags. His fully packed bags held about 250 abalone (80 kg), and he liked the deckie to send down a bag every 15 minutes. He used nitrox because he maintained it was safer and healthier. However, others in the industry reckon that rather than making his dive profile safer, it stretched them out.

In the end he embraced modern diving and was considered one of the Zone's good divers.

Blue Grant passed away in January 2008. *Based on an interview with Robert Coffey, 28 May 2003.*

John Hollingworth
Aba-what?

John Hollingworth was only in the industry for a couple of years in the mid 1960s, but he still reflects on how different his life would have been if he'd applied for a licence when he had the chance.

'Every man and his dog got his licence for two quid [£2 or $4] and they told me I'd better get one. I thought I'd just walk in and put my two quid down and get it. But, being a bit lackadaisical, it took me a while before I finally went to Fisheries. By then they'd shut the door and haven't opened it since. I applied and reapplied. Then the price went up and up and up for a licence. I was devastated.'

John joined a spearfishing club at Warrnambool, which is where he met Andrew Coffey. One day they were having lunch and John recalls Andrew saying, 'I've got an idea about how to make money.' John says this was typical of Andrew — always coming up with bright ideas. Andrew said, 'We'll get abalone and sell them.' Very little was known about abalone then and the man sitting next to them said, 'Who the hell would buy abalone?' John was given the job of finding

potential buyers for the abalone because, as far as they knew, there was no one in the area selling them. 'I had to hunt around to find a place to get rid of them. I went to all the wholesalers in Geelong. When I told them of our plan, I got a lot of "Aba-what? No, mate".'

John was running out of people to contact. Then he went to Blackney's in Geelong and they said they'd see what they could do and to come back in a month. When John returned he was told they could 'dispose of them'. Although that was good news, there were stipulations. The abalone had to have a lot of salt in them as well as their juice. They also had to be shucked. John took the news back to Andrew Coffey and that's how his and Andrew's involvement in the industry began. It didn't take long for the others to come into the industry until there were around 500 men diving, before the introduction of licences.

'Andy and Don Barker got their licences. I didn't know you had to have a licence because they were just like snails in a garden. No one looked at them. They were just there like vermin. No one in Australia would eat them then. At Warrnambool there were layers of stone shelves like a library and there was just so many. So many you didn't even need a tool to get them off, just a screwdriver. It was unbelievable.' John recalls getting £2 8s ($5–$6) per lb, which was a lot of money in those days.

John spent most of his time diving in Port Fairy and recalls how they made all their tools of trade. Most divers went out fishing in boats, but 'Andy [Coffey] had the gift of the gab and talked his way into getting beach access from all those properties along the coast. He used to drive down to the beach with his tinny on the back and launch from there. While most of the other divers went out in boats, he used an inflated rubber tyre tube and a potato sack with a rope and a reef anchor as his tools of trade. It was all free diving in those days and we would get maybe six abs at a time. Andy would paddle out from shore to the lagoons, islands and reefs. He was like a fish. He'd always get more than anyone else. The others thought this way of fishing was a bit of a joke until they found out he was getting more than they were. He was a master.'

John says Andrew sometimes would go to his other job at the local newspaper still in his wetsuit, because he didn't have time to go home and change.

Andrew and John used to store the abalone in potato bags or sugar bags. The bags were so heavy they couldn't lift them and so they dragged them to shore. John remembers Don Barker being a 'wonderful bloke' who was so big and strong he could walk out of the water with a sugar bag full of abalone over his shoulder. Don would walk the kilometre or so up the cliff to the car with the shucked abalone in a bag on his shoulder. All the froth from the abalone would cover his body and be oozing down his legs.

Andrew Coffey went on to invent a stretcher-type carrier with a rubbish bin in it to carry the abalone.

John says he used to worry about carting the shucked abalone in plastic bags as the bags had been used and they had no idea what had been in them. And the bags would leak. After the abalone had been out of the water for a while, they would smell — they weren't off, just smelly. John's daughters Julie and Kay used to do the shucking and John says Julie was always complaining about how she was forced to do the work.

He recalls one day Andrew was diving a long way out when they saw a Fisheries inspector coming towards them. Andrew surfaced and had a long, friendly chat with the inspector. John kept a low profile and didn't get involved in the conversation, but as the inspector was leaving he suddenly turned to John, pointed his finger at him and said sternly, 'Listen, mate, if you'd come out with an ab bar I would have got you, for sure.'

As John remembers it, 'They were good days and I used to love the life because we were all young, fit and determined. Wonderful days. Once I couldn't get a licence it all changed because I couldn't work with Andy.'

John lives in Breamlea, near Geelong.

Dick Kelly

Tracksuit Dick

According to Lenny Burton, Dick Kelly's nickname was 'Tracksuit Dick' because he always wore a tracksuit. 'He was a good bloke, but he was a strange bloke,' he says. 'I know he was strange because he boarded with us for a while. He was a big, strong bloke who'd fought in the Korean War. He was in the air force distributing propaganda pamphlets into North Korea and was shot down. He parachuted out and he was stuck in a tree in his parachute for three days, right on the line between South and North Korea, the demarcation line. He hung in that tree until he was found, and he didn't know whether it was by the North Koreans or the South Koreans. He was in a mental institution for a long time after that. Then he recuperated and was driving ambulances for a while in Sydney before he came down here.'

Lenny answered an advertisement in *The Age* for someone to go abalone diving in partnership with Dick. Although Lenny didn't have a lot of diving experience, they were a perfect match — Dick had a boat and Lenny had a bit of gear. They went to Blackney's at Geelong and asked where the best place was to find abalone and they were told to go to Torquay, because they were getting too much blacklip and the market wanted greenlip. 'Everyone had been to Torquay and fished the place out. We didn't know that so we were the only two divers there.'

Dick had a 17-foot (5.18-m) De Havilland Hercules, an aluminium boat with rivets, and they worked Torquay for a week. It took them about four days to find their first abalone, because they didn't know what they were looking for. Then, as they were swimming around they saw a rock that moved and it was an abalone. 'Once we knew what we were looking for we were right. We only made $20 each the first week so we weren't doing too well.'

Then they discovered that everybody was at Portland. When Len and Dick got there, about 60 boats were already there.

Lenny left his wife in Melbourne and he and Dick camped at what was then the Centenary Caravan Park. They would come home after diving and Dick would sleep in the back of his van. 'He had a big Ford F250 and I'd sleep in the auto tent. We had a little TV in the auto tent and we'd sit there with no heating or anything, wrapped in blankets, watching this little television and eating crayfish that we'd got that day. But you get sick of crayfish after a while.'

Lenny recalls an incident when Dick was diving at Lady Julia Percy Island. His sheller saw two shark fins converging on Dick while he was in a bit close to shore. The sharks were working in unison. They went straight in but they attacked a seal that was next to Dick. 'He never knew anything about it. Dick just kept on working. His sheller pulled him in, but Dick didn't see anything. Those sharks had come from way out and were zeroing in on this seal. They would have known Dick was there too, but fortunately they picked the seal.'

Frank Zeigler, ex-police officer and dive instructor, remembers Dick telling him he bought his original fishing licence for a shilling and that he grumbled when a couple of years later it went up to a £1 ($2).

Frank often used Dick as a 'test pilot' for trialling new dive products and procedures, because he was willing to try new things. If it worked, Dick would tell the other divers about it and they'd try it. He tried things like breathing pure oxygen on the way home from diving, wearing a depth gauge and using proper dive plans. Frank says Dick would keep a log to make sure he knew what was going on. He had a thirst for information.

According to Frank, Dick was one of those guys who actually planned an exit strategy from the diving industry. He always said when he got to a certain point he'd sell his licence and start making films because he didn't want to dive forever. He'd got his lifestyle and he was happy. Unfortunately, Dick died of a heart attack several years after he sold his licence. He had a lot of health issues and it's not known if these were connected to diving.

Bob Hope adds, 'Dick was always in a tracksuit even though he had so much money it wasn't funny. He was rich even before he started diving. He had rich aunties in Sydney and when they died they left him an inheritance. But after the Korean War, he was scarred for life, which might explain the way he behaved sometimes.'

Joe Milani, managing director of Southern Canning, says Dick was one of the few divers he was friends with. According to Joe, Dick was an introvert and didn't mix with many people. On reflection, Joe thinks he could have been one of the few people Dick trusted. 'Perhaps I attract oddballs, the ones who don't associate with other people.'

Noel Middlecoat recalls the tragic incident when Dick and his deckhand Ronnie Baird were out at sea when a big wave came and they realised they had to get out in a hurry. 'Ronnie fell out of the front of the boat and virtually got decapitated by the outboard motor. It was a Shark Cat catamaran and he got caught between the two hulls. He was a good mate of mine.'

Noel Middlecoat

A few dustbins full of abs

Noel Middlecoat was one of the divers who was there in the fledgling years of the industry and spent time in the Western Zone, before electing to take up a licence in the Central Zone. He was in Portland between 1967 and 1968 and then 1974 to 1975.

'I worked the whole coast, from Wilsons Promontory to Portland, so I had a choice about what zone I wanted to be in. Some of the older divers didn't want zoning but the newcomers thought it would be good because they wouldn't have to travel up and down the coast. The married ones wanted to stay put and the single ones wanted to cruise up and down. That was a pretty good lifestyle,' he says of those early days.

Noel maintains that the money didn't really come into it until the licences became saleable. For him it was all about lifestyle. 'Then all of a sudden an abalone licence was worth millions of dollars. A lot of early blokes like myself, who got out before that big jump, didn't make the money that they've been making in the last few years.'

Rick Harris and Noel began diving for abalone at the same time. They were both plasterers and mad keen spear fishermen. Noel's father was a scallop fisherman and told him about what he called 'muttonfish': 'They'd pull up their dredges of scallops and there'd be rocks with abalone stuck all over them. So we went and got a few.' The only problem was that the men didn't know where to sell the fish.

Noel went to New Zealand and when he came back Rick told him they'd found a market. 'We took a few dustbins full of abs and they bought them off us. We probably got around 10 cents a kilo.' Noel thinks there were probably 10 abalone boats in Portland at the time.

According to Noel, Portland was then a 'funny sort of town. It was like an old wild west frontier town because, along with the ab divers and professional fishermen, there were a lot of meatworkers and riggers who came from Melbourne. The pubs were always packed and there was a lot of drinking going on. Then when the weather came good they'd go and do some fishing.'

In an industry where knowledge about good patches is tightly held, diving techniques are guarded and there's lots of money to be made, there will invariably be jealousies among the divers — and there were.

According to Noel, when the Mallacoota divers moved into Portland they were seen by local divers to be taking over, especially when it came to using valuable space in the freezers they used behind one of the cray fishermen's yards. 'We didn't mind them coming here so much, but they tried to rule the roost. You know, take over the freezer.' He says the mood among the local divers was 'Don't let that bloke take over or we'll do this and we'll do that.'

Noel was also acutely aware of tensions between the married and the unmarried men. He says he used to go to New South Wales in winter with Bernie Morton and Jurgen Braun to escape the cold weather in Portland. They all had New South Wales licences. 'A lot of the married blokes couldn't do that and were jealous of our lifestyle — taking off in the depths of winter and living in beautiful Narooma and Tathra. They didn't like that. When we got back, they got funny on us because we'd lived like nomads up and down the coast and they couldn't. That's when they started making a fuss about having a special Western Zone.'

Noel recalls the tensions about these types of issues. One day he went to the local petrol station to get fuel for his boat. Noel knew the station proprietor, Ronnie Baird, who was a deckhand for Dick Kelly. (Ronnie was tragically killed at sea when a huge wave swamped Dick's boat) Ronnie said he'd been told not to sell Noel petrol because he was one of the lifestyle divers who left town for the winter. 'When I asked him where I stood about getting petrol, he said, "I told them to go and get lost and that no one tells me who I sell petrol to and who I don't sell petrol to." Then he served me petrol.'

There was a similar incident with the fish carter who used to handle the divers' abalone. One day the man approached Noel and said he'd been warned that if he carted the catches of Jurgen, Bernie or himself, he could not handle the catch of the others. 'It was like a dozen divers against three. I told him that it was his business at stake and that we'd cart our own fish. So Yogi [Jurgen] bought a trailer and we used to take it in turns to take our own ab to the market. There were a lot of things like this — internal fighting, in the early days.'

There was the time Johnny Munz got badly bent. 'There were some rig divers in town and they got a decompression chart from their rig and gave it to us. We went and got him out of hospital and took him out to sea. We took him down to about 100 feet [30 m] and slowly decompressed him. I was the diver who went down with him and it took us maybe three or four hours. I had divers coming down with hand signals telling me how many stops we had to go up. When we came to the last stop about 10 to 15 feet [3–4.5 m], we had to stay there for 25 minutes. It was pitch black and Portland was a couple of kilometres away. The next day he was a bit sick, but he turned out to be okay. I saw him years later in Melbourne and he never had a problem from that attack of the bends.'

The only time Noel had the bends was at Wilsons Promontory. He couldn't dive after that for 12 months.

Bernie Morton
Shark-proof in his footy jumper

Bernie Morton and Jurgen Braun were dive partners for a while. Bernie was the twin son of Tex Morton, the New Zealand country and western singer.

Jurgen first met Bernie in Eden in 1966 when they were both working on the boat owned by Tony Jones. Ivor Magnussen was the skipper and he used to get a percentage from each diver. After two days' diving, Ivor's percentage was enough for him to lift the anchor and go home. Bernie decided he couldn't earn enough money working like that so he returned to Sydney. At that stage the price of abalone was low and he told Jurgen to contact him when it improved and he'd come back again. And he did.

Jurgen had moved to Mallacoota and, when the price went up, Bernie visited for the weekend. He said that if he couldn't make enough money diving he'd go back to his steady job. The weather was rough so they fished off Bastian Point, outside the Mallacoota bar, where they could launch off the beach. Bernie earned more money fishing on the Saturday and Sunday than he did in a whole week as a brickie's labourer. He went back to Sydney, gave a week's notice and came back to Victoria for good.

Bernie had the boat and Jurgen had the fishing-ground knowledge. They pooled their money and six months later had enough money to buy a bigger boat. 'That's how we started,' Jurgen says. 'Bernie was good for me. His mum and dad had split up when he was

young and he was always into saving money to buy flats or any sort of investment. He helped me a lot. I paid him back and helped him out at one stage when he had Hepatitis B and wasn't able to dive for six months. I said to him, "We're in a partnership and we split everything in half." So I paid him back for helping me get into the abalone industry.'

They managed to put together $1000 as a down payment on an 18-foot (6-m) fibreglass Bertram. They worked it from August 1967 but did a drive shaft in the outboard after a couple of months. Their boat was being repaired in Melbourne and they thought they'd pick it up and check out the west coast, because the grounds at Mallacoota were becoming depleted. They saw bluestone capes at Portland and it looked fantastic, so they decided to dive there. They found some good abalone. 'We sneaked back to Mallacoota, stayed another few days and we packed up and left in the middle of the night without saying a word to anyone.'

That was in December 1967. In Portland, Bernie and Jurgen bought a cray boat and went cray fishing as well. At the beginning it paid about the same but it just wasn't as exciting as abalone diving. Jurgen was in partnership with Bernie until 1973, when he moved to Port Fairy while Bernie stayed in Portland.

Jurgan remembers Bernie wearing a football jumper when they were diving off Lady Julia Percy Island because he thought the other divers looked too much like seals. Bernie would say, 'When a shark comes along, we're the slowest things and we look like seals — and that's what they'll go for.' But Jurgen has doubts about whether the jumper actually worked.

Like most of the divers and deckhands who were around in the early days, Kim Heaver has many fond memories of Bernie and recalls him wearing his legendary football jumper. 'A lot of blokes didn't get along with him, but I always did. One night we were coming back from Lady Julia Percy Island and we ran out of fuel about halfway. We let off a flare and Bernie came out in the dark and towed us back.'

And there was the infamous stolen boat story. Those involved have different takes on the incident that's become folklore in the Zone. One story goes that the Portland divers agreed to voluntarily close off inside Nelson and reopen it at a certain date.

Apparently Bernie went there the week before, which upset some of the others who took Bernie's boat and dropped it somewhere without damaging it. But Bernie had two boats so he went out in the other one until the missing boat was found. One version of this tale goes that it wasn't found until years later; another says it was found a few months later when a bushwalker came across it in the bush.

Tony Jones remembers the story a little differently. 'We took his boat and hid it in the forest. I was the instigator because I felt I'd started him off in the business. I said if we're going to do something, then we have to do something that affects his hip pocket. We went to his house at about 3 am and physically picked his boat up. We hooked it onto our trailer and hid it in the bush. He knew we'd done it. When the police came to us, we said he was a bit of a rogue and they said to sort it out between us. About seven months later a guy was walking in the bush and came across the boat.'

Bob Ussher, who remembers Bernie as a good bloke and a good friend, recalls one time Bernie got stuck on a sand bar at Nelson. His wife rang Bob at midnight saying that Bernie wasn't home and could he go and look for him. Bob went out to Nelson but couldn't budge Bernie's boat from the sand bar. 'We got him off it about 2 am and rowed him up the river. He got home about 4 am, completely buggered.'

Peter Riddle
A true hunter-gatherer

Peter Riddle is one of the younger divers in the Zone. He started spearfishing when he

was a kid in New South Wales. 'I fell in love with the ocean and all I ever did was dive.' He used to fish, longlining, for gummy shark but always wanted to try abalone diving. He got to know a licence owner. 'I was selling crayfish at the same place he was selling his abalone,' Peter explains. 'I got to know him over the years and, because I could dive for crayfish on my fishing licence, I caught as many as I could to make it look like I was a good diver.'

After four years of abalone diving in New South Wales on Gary Allen's licence, Gary bought a licence in Port Fairy from Dick Cullenward and asked Peter if he would consider moving to the Western Zone. Peter began diving in Port Fairy at the start of the season in 1998.

These days he dives several licences. 'I've always caught extra quota,' Peter explains, 'and been on different licences. I've even dived for a licence in the Central Zone a couple of years ago. You never know who will ring you and ask you to dive for them.'

For Peter, the good things about diving are the calm days and good catches. A good catch is a tonne a day but he doesn't aim for that. Rather, he usually aims for 500 kg if he can get it. But it depends on the area he's diving in. 'A calm, clear sea is what I really enjoy. I bought my house so I could stand here and check out the ocean from the window — I don't have to drive anywhere. I can tell exactly what swell there is from home because I've been looking at the sea all my life.'

He is mates with the other divers but admits it's a secretive job because there are unwritten rules: the divers have to be the first to get to a good spot and then they have to keep 100 metres apart so they don't encroach on where another diver is working.

'People know what you've got because they see where you've worked and then that spot's lost. That's why it's a secretive job. You have to play cat and mouse. You park well away from a spot and get changed and see if anyone comes along. Or you leave late so you know everyone has gone to work and then go to your spot. You try to keep it to yourself but it's a bit of a game. If you don't get seen on the day, then you don't get found out. You can come in with a big catch and as long as no one's gone past and seen where you are, you've got away with it. But eventually someone will find out.'

Peter dives both shallow and deep, depending on where he is and where the better abalone are. It's a dangerous job, he says, because the diver gets air pumped down to him through a line and he can drown if the boat moves suddenly and cuts out or he loses the regulator from his mouth.

Peter used to dive 'live', leaving the boat free from the anchor, and the deckhand follows the diver around in reverse. 'When I first came to Port Fairy in 1998 I continued to use the same live diving system and quickly realised that I was the only one doing so. All the other divers were working anchored up. As my boat was always in the reverse position to their boats, I was receiving some very strange looks. On one occasion another ab diving boat came over to where I was working and drove around my boat, staring and wondering what we were doing. It then drove off.' While many divers thought live diving was a dangerous practice, now all the Western Zone divers work live.

'You really have to trust your deckie,' Peter explains. 'Your life is in their hands. It's not a normal job to spend all day under the ocean. If I dive deep one day, I won't dive deep the next or vice versa. If some areas have lots of deep spots, I don't do two days

Clockwise from top: Rob Torelli searching for abalone through a seaweed bed. Peter Riddle with sealed bins of abalone. Catch on Peter Riddle's boat. *Photo source*: Rob Torelli.

in a row. I'll do a day, have a day off and do another day just so I won't get the bends.' He's had the bends three times. He now carries pure oxygen on his boat, which helps him to avoid the bends.

Killer whales have swum around Peter. 'You get dolphins but it's rare to see a whale,' he says. 'I've never seen a white pointer but I've seen other types of sharks and actually touched them. They usually appear when you least expect it — all of a sudden they're there.'

Peter has had his boat completely fill with water. 'I'd copped a wave. I had put it in a bad spot, my own fault. You can get huge seas, so you have to pick your weather. Sometimes when you're catching a lot of abalone, the deckhand will back into it. They work backing into it all the time, so the waves are coming over the back of the boat. If you're catching a lot of abalone the deckie will hang in there as long as they can. You come up and they're drenched.'

Peter says that divers can get a bit brain dead at the end of a day's diving, but then they've still got heaps of paperwork to fill in for the Fisheries Department. He has to fill in log books, and phone through information regarding the different reefs he worked and which of the three licences he was working on that day.

When he first came to Port Fairy, the divers could dive wherever they liked but had to tag the abalone. Now all the bins have got to be sealed and each reef code has to be written down and phoned through. Peter has a machine on the boat that measures each abalone, gives it a GPS mark and the catch rate. 'It's called a data logger. They've only been around for the last four years.'

According to Peter, the running costs and the distances travelled now are huge compared with what he used to do and where he is able to dive. 'It's a lot bigger cost on me to catch it. We have longer boat rides — one of them is 100 km to a reef where we dive on the South Australian border. I've got to drive to Portland, which is an hour, then I've got a two-hour, 100-km boat ride to catch a dismal amount of abalone, probably 300 kg. You leave at 5 am and you get back at 6 pm. You have to know the weather is going to be good in that spot before you set off.'

Peter eats abalone very occasionally, only cooking it if someone asks to try it. He says you've got to tenderise it first, then you can do whatever you like with it. He likes people to try the abalone with little else — maybe stir-fried with vegetables and a simple sauce. Or whole crumbed abalone pan-fried with lemon juice added at the end. The best way he knows is to leave it whole, give it a whack with a mallet, not too hard but just enough to break the sinew, and then it will be tender. 'There's nothing worse than when it's chewy. If it's tough it's terrible.'

When he is not diving, Peter cuts firewood. 'That's going really well for me. If I stopped diving, that's what I would probably do.'

Peter Ronald
A good life? It's been a boy's own adventure

Peter started diving in 1992 and he worked around 40 days a year. In that time he made more money than he could ever imagine, even as a lease diver. 'Only 40 days a year and I'm messing about in boats and four-wheel drives, going places, seeing the ocean floor that no one's seen before, killer whales out at Julia Percy or seals or whatever. It's been fantastic, absolutely fantastic.'

Peter came from a family of swimmers and his love of the sea began when he was four. He has an interest in maritime history, which led to diving on shipwrecks in the Peterborough and Port Campbell areas from about 1970. He has long been involved with Flagstaff Maritime Museum in Warrnambool, becoming its director. He held that position in 1992 when he started abalone diving in his spare time at weekends or holidays. Eventually he resigned from the museum and was abalone diving full time.

At the time Peter became involved in the industry, he thinks there were probably three lease divers and the rest were owner divers. 'It was a really interesting group, an amazing mix of people like Phil Sawyer and Konrad Beinssen. Phil was an intelligent guy but somewhat eccentric.'

Peter began diving on Dick Cullenward's licence. Peter fondly refers to Dick as the 'Mad Yank'. Dick was the first in the Zone to have a nominated diver — Donny Ray, who had been Dick's deckhand and was his first diver. When Peter began diving for Dick, he took over not only his licence but also his usual fishing patches. Dick took him to Lady Julia Percy Island with a video camera and showed him where to dive, and Peter soon fell into Dick's pattern of diving.

In the late 1980s Peter recalls that Dick

went to a boat show in Western Australia where he spotted an aluminium-hull cat cruiser complete with toilets, showers, bunks, galleys and a hot water service. He purchased it and brought it back to Victoria and that became his abalone boat. The *New Toy* was an amazing boat, recalls Peter, about twice as big as anything else in the area and completely over the top for an abalone boat. 'Dick was being the extravagant American. The boat was great for diving out of Julia Percy Island. For that kind of anchor diving it was perfect — like a floating hotel.'

Peter says there was an unwritten code that you had your patch and everyone knew that and respected it. 'Robert Coffey and I dived at Julia Percy almost exclusively in the first 10 years. He had his areas and I had mine. If he strayed into my patch, then I would do a pay back and stray into his. That was how order was maintained. But it wasn't something you'd look to do, because your aim was stability and order, so it was kind of a natural process. It was never spoken about, never discussed. It was just the way it was done.'

Peter remembers many colourful characters in the industry. He tells the story of being at a high-level meeting with the head of Fisheries Enforcement, when one of the divers came in and lit up a joint. 'The enforcement head didn't know where to look or what to say.'

When Peter became secretary of WADA in 1993–94, one of the things he had to get used to was the clash of different personalities. 'You'd have the committee meeting and someone would come up with what I thought was a pretty good idea, but other people would say, "Oh, that's rubbish, you can't do that".' He soon realised that when some of the guys heard an idea the first thing they'd ask was who was making the suggestion. If it was someone they clashed with, they would vote against it. He says he had to get a handle on how these politics were unfolding in terms of how people would respond to new ideas. 'That meant a bit of creativity in how things were presented and got through the committee. There was no pretence at logic. It was blatant.'

Funny things happen below the water. One day, Peter says, a seal was watching him, then it put its head in his bag, pulled an abalone out and dropped it on the bottom in front of him. Peter picked up the abalone, put it back in the bag and the seal did it again,

playing with him. 'They don't usually eat abalone. That's about the only time I've seen a seal carry one in its mouth.'

Peter felt on edge whenever he saw a seven-gill shark, a common coastal shark, which is a bit like a tiger shark. They're not regarded as being dangerous, but they have the kind of teeth that enable them to take a bite out of something that's bigger than they are. Peter says there used to be a charter boat for recreational fishing working out of Warrnambool. They'd take people out on the boat and the entertainment was to catch a seven-gill shark, pull it alongside the boat and leave it in the water so the other seven-gill sharks would come and eat it.

'They're cannibalistic, opportunistic feeders. So they're certainly dangerous. They're excited by the scent of the abalone trail, which is why they're coming in and they're attracted by food. They're hungry and they're looking for something to eat. If I had an abalone tool, I'd hit them with the tool to try and get rid of them. That was generally effective. The last time I hit one with the tool, its tail flicked me as it went past. The sweep of its tail hit me in the side of the head, not badly, but it was that close.'

Peter recalls Phil Plummer having an experience at The Crags, the most common place to see the seven-gill shark. Phil was swimming through weed and felt some resistance on the parachute of his bag. Phil looked around and there was a seven-gill shark with a plastic bag hanging out of its mouth. It was before the days of the electronic shark repellents and the shark had taken a bite out of the bag.

There's a huge seal colony at Lady Julia Percy Island. 'In those early days, the seals would pester you just like playful dogs. They'd bite the top of your wetsuit hood, then pull it up and let it go. When they pulled it up, it would fill your hood with cold water and when they let it go, a jet of cold water would shoot down the back of your neck.'

One day at Julia Percy, when he was anchored on the north side having a cup of tea, a seal got up on the boat's back platform and waddled up to the cabin at the front. The deckhand and Peter were amazed to see it look into the cabin of the boat, walk back, perch itself on one of the motors and sit there looking at them. It had no fear.

There used to be huge stingrays in the early days at Propeller Bay on Lady Julia Percy Island. Attracted by the smell of the abalone, they would swim over the top of the diver, literally covering him like a blanket.

Peter would hold an abalone up, stick it in the stingray's mouth and it would swim away and come back with its mates for more. 'They were harmless — just looking for a feed. That was a highlight of my diving at Julia Percy.'

There was a code among divers that you would never work within a hose length of another boat. But by the time Peter stopped diving, that code was forgotten because there was so much competition. 'On a busy day at Julia Percy, there'd be three boats diving around the Island. Last year when I went out for the opening of the season there, there were eight to 10 boats in Propeller Bay alone. What that means is there's no stone unturned. It's only a small bay, about 800 metres long, but those boats were there for three, four or five hours. They're totally scouring the bottom to remove any fish of legal size. That's okay if it is legal size, but it is a huge shift compared with what I was used to.'

When Dick was showing Peter his patches at the Island, he'd say, 'See that rock there, you can go there tomorrow and get 400 kg of abalone. Leave it alone for a year, come back and you can do the same again.' For eight or 10 years Peter was able to do pretty much that, but things started to change. 'What I now know is that because of the collapse of some major reef codes, all of a sudden it became more competitive. Where I'd had a spot to myself for years, now there were other divers fishing in my areas and creating competition. I now know this is a common symptom of a fishery in trouble.'

The best story about cooking abalone he heard from a Fisheries minister who was speaking at a conference in Adelaide. 'He said that when he was at a processing plant in Adelaide, he asked one of the workers, "How do you get a decent feed out of one of the things?" The worker replied, "You sell them and then buy some steak".'

Peter has his own method. 'You wrap them in a tea towel, one at a time, and belt them with a mallet until the rigor mortis is gone. You can feel the change. Once they become pliable. then you can do what you like with them. Cook them whole on a barbeque, slice them, but the real trick is getting them tender before you cook them.'

He stopped diving in December 2006.

Rob Torelli

Peace and happiness underwater

Rob Torelli has never had a 'normal' job. He's an underwater cameraman, underwater film-maker, recreational diver, charter boat operator and a seven times Australian spearfishing champion.

He also dives four abalone licences in the Western Zone, where he's been diving since 2007. He has the most diving quota commitment across several licences. This means he spends around 100 days a year diving, which is a big commitment in this industry. 'I just love it. I feel far more comfortable underwater than on land. I find more happiness and peace when I'm under and that's why I spend so much time there. I switch off the moment I jump in.'

Rob got into the abalone industry through his brother Richard, who was the managing director of Tasmanian Seafoods for 17 years. Richard was a former abalone diver, as was their brother David. Rob tried getting a job diving for abalone in the Central Zone but found it difficult to get work there. 'I've dived for sea urchins in New South Wales and work underwater through my film-making and do dive charters. I had a good

boat and, when the opportunity came up in the Western Zone, I took it. I've done other work in the Zone so know the waters.'

Many divers have their own techniques for diving — things that work for them. For Rob, it's nitrox. He believes nitrox means less chance of getting the bends and less nitrogen in his bloodstream. Rather than having a normal mix of air and nitrogen, he has a 40 per cent oxygen mix. He's set his boat up to accommodate the 5- to 6-foot (1.5–1.8 m) nitrox bottles he uses. 'It's not for everyone, but I need to dive deep on so many consecutive days. I get good scores at around 25 metres and dive to 30 metres, whereas a lot of divers stop at 20 metres. Nitrogen builds up in your system and you can succumb to the bends unless you address it properly. Nitrox allows me to dive deeper and for longer than the guys on hookahs.'

As well, Rob does 100 per cent oxygen decompression underwater, a method he believes is three or four times better than doing it on deck. When he surfaces, his deckhand gives him an oxygen regulator and hose. Ironically, while this system is used to keep the bends at bay, if the diver goes past 8 to 10 metres, he is likely to black out and die. Rob makes sure he has a short hose and regulator in a different colour, so he can't take these to the depths where he'll black out.

Lady Julia Percy Island holds a special mystique for Rob. 'I love the seals. I love the cleaner water. While a lot of the guys have a thing about deep water, I love it.' Although he's seen a few seven-gill sharks, he's never encountered a great white while abalone diving — only when he has expected to see one on his charters and when underwater filming. However, he knows there's a good chance he'll see one out there one day.

In 2011 Rob hosted a French film crew who made a documentary on his work as an abalone diver for the long-running French television program *Thalassa*, which was shown in Australia on SBS. During the production he worked with the host and introduced her and viewers to abalone diving, including filming at Lady Julia Percy Island.

Rob lives outside the Western Zone and commutes to work. He's good mates with Peter Riddle and they share a love of spearfishing. He says Peter is a true 'hunter and gatherer'. According to Rob, 'If every diver was a good as him, the industry would be in good hands.'

Diving on multiple licences in the Zone meant other divers lost their jobs and Rob recalls some being upset at the time. However, these days he is happy to have a good working relationship with the other divers.

Rob's secret to cooking abalone is to keep them tender. 'I always keep them live in the shell until just before I cook them, and then it's on a really hot barbeque. Rigor mortis sets in once they leave the shell and they go tough, so you've got to tenderise them properly. First I shuck them and put them in a towel or T-shirt, or something where they won't slip. I then hit them with a bit of four-by-two or a brick. Only one or two hits. This shatters the muscle without pulverising it. Then they go on a hotplate for a couple of minutes before I cut them into bite-size pieces and serve with dipping sauces like soy or wasabi.'

Daniel Ussher
Divers need a healthy disrespect for their own well-being

Bob Ussher is a hard act to follow, but his son Daniel has done his fair share of diving and come away from the experience with a contemporary and realistic understanding of the challenges and dangers divers face.

Daniel's earliest memories of the industry date from when he was around 12 years old and there was a big drop in the market for abalone. There were also problems with pricing and getting the abalone to market. Daniel remembers his father coming home after a day's diving and helping to hook up the tandem trailer and loading it with abalone. 'I'd hose the ab down, put hessian bags over them and then Mum and I would drive them to market in Melbourne. I think I was there to keep her awake because we wouldn't get home until the early hours of the morning.' Bob would go out diving the next day and Daniel thinks they did that for around three months because there was no processor in Portland. Daniel says they sold the abalone to two men who had just starting processing — Lou and Danny with a surname like Dalozzo. 'These guys were really loyal to Dad. They said they'd pay him a good price and he'd be paid as soon as they sold the abalone. They often said how he'd given them a start in the industry.'

Although he spent a lot of time snorkelling and surfing when he was young,

Daniel didn't have much interest in the abalone industry. He was around 25 and in Brisbane when Bob broke several ribs and hurt his back in an accident. 'Dad phoned and asked if I'd like to go diving. I was down there in three days ready to start.' Bob's back was never the same and for around four years Daniel dived full time on Bob's licence and using his boat, fondly known as *Bertha*. *Bertha* was eventually sold to David Forbes and became *Absession,* which sank.

Daniel reckons he inherited the Ussher trait of stubbornness. He broke his ankle in a motorbike accident in his second year of diving and it was put in a cast. But he still had 800 kg of quota to catch and only three weeks to catch it in. While they could have sold the quota, Daniel vowed he could do it. Although it was painful and difficult, he managed to dive by putting a tyre tube over his leg and taping the end to stop the plaster getting wet.

Big catches are often talked about. Daniel caught 1400 kg one day. Jamie Espie and David Forbes reckoned it was the biggest catch in the region, but about three months later one of them came back with 1500 to 1600 kg. And then a few months later one had a bigger catch again. 'It was on to see who could catch the most.'

Daniel has a very philosophical approach to diving and believes the thing that draws men to the job is the escape they achieve when they go below the water. He says divers know the consequences of every decision they make underwater. 'When you're diving there are no mobile phones. No thinking about what you did last week. You only focus on the job. You escape because you have to focus so much on what you're doing in terms of safety. It's profound and very few can do it.'

From the moment they put on their diving gear and drop over the side, it's a serious business. 'Divers enter a different zone and become a different person. They are self-driven and don't like to be told what to do. Every choice they make is life or death. Every choice is important and has consequences. I struggle with this — most people simply don't get that.'

Most divers have had the bends and Daniel says he struggled to get back into the water after his experience, after which he started having doubts and fears. 'It's cold. It's wet and miserable. You might not have

dived for 30 days and everything seems to be against you. But you have to push through that barrier to keep on going.'

The pain Daniel felt when he had the bends was the worst pain he's ever experienced. His boat's propeller cut his airline and, while he had a 6-litre bail-out cylinder, that's only good for about 15 minutes coming up from depth and not enough to decompress. Within minutes of reaching the surface he felt pain and within 15 minutes was screaming in agony. Daniel remembers Bob looking at him and saying, 'It's bad.' Those words were not a good sign from someone who'd been in the chamber as many times as Bob had.

At the time Frank Zeigler had just finished building a recompression chamber in Portland and even though it hadn't been trialled, Bob arranged for Daniel to use it. 'I was in the chamber for eight hours and could only just walk when I got out.'

Daniel thinks Bob was in a chamber anywhere between four and seven times and there are a host of stories about his experiences. So much so that, after perhaps his fourth time in the chamber, the National Safety Council gave him their book *The*

Diver's Medical Companion and advised him to read it.

One time after getting the bends, Bob was put in an ambulance to Adelaide and, to save time, a portable chamber was rigged up on the back of a truck to meet the ambulance en route. However, the truck got stuck under a bridge and was jammed there for four hours. Another time Bob was sent in a low-flying aircraft to decompress in Adelaide. He was put in a portable chamber on one occasion and said it was the most horrible experience he'd ever been through because it was like a coffin.

Bob suffered a near-fatal air embolism (air bubbles in the vascular system) and it took them eight hours to get him into a chamber. He came out of it in a wheelchair. Afterwards the medicos did tests on him and told him they thought he had suffered brain damage from the embolism because he was a little 'slow' and stuttered. Bob said to them, 'Why do you say that? I speak better than I did two years ago.'

Daniel comments that 'Dad had bad bone necrosis when he finished diving but three years later they did a bone scan and it had completely healed. They reckon it was because he was diving continuously and diving outside the tables. After a few years out of the water, the nitrogen gradually worked its way out and the blood flow came back. The bone healed itself.'

Bob also had his fair share of shark experiences. One day he saw three white pointer sharks at three different locations. Another day a white pointer came straight for Bob and hit his bag, an incident that disturbed him so much that he lost his nerve and stayed out of the water for months. Daniel remembers Bob saying that, 'It scared the shit out of me.' On another occasion Bob was circled by three sharks and sat on the bottom for three hours until they went away.

Daniel says his father was a hard worker who used to work when others didn't. 'He had a constant need to make money and to work hard to provide us with things he never had, which he did. But it meant he didn't spend as much time with us as he would have liked.' One time the family was one week into a two-week holiday in Bali when Daniel became sick and his sister had a fall. Bob announced that because they were sick and the weather had come good, they were going home so he could go to work — and they did.

Reflecting on what drove his father to work at such a frenzied level, Daniel says it went back to his tough childhood. 'I'm not sure Dad even finished primary school, but he had an abusive father who walked out on the family when Bob was young and never came back. Dad was forced to earn money to keep food on the table.'

Daniel is no longer diving but might return when the quota picks up — and he firmly believes it will.

Glenn Plummer

It's in his genes

Glenn was always destined to be an abalone diver — it's in his genes. Dad Lou holds a number of licences, and Glenn and his brother Phillip fish on one licence each. Glenn's son Brendan is a casual deckhand and, just to keep it in the family, his wife Camille even did a stint as a deckhand in 2004. Glenn is also a major shareholder in the abalone farm at Port Fairy: Southern Ocean Mariculture.

After studying at university and taking up a job with the Office of Corrections Glenn decided he wanted to get into the abalone industry. He began working as a deckhand for Lou and also for Phillip in 1995, and is among the new breed of divers.

But, as Glenn acknowledges, it's a tough way to earn a living. The work is sporadic and he can only work when the conditions are suitable and he still has quota to catch. Because the work is on–off, when he goes out he spends seven hours straight underwater to maximise the conditions. The downside is that his dive days can be anywhere between a day to three months apart. 'You just keep on going at the time. But it's when you stop that you feel it. It's usually on the second day after diving that your body crashes. It's not the diving, but the breaks that do it to you.'

Glenn says divers must have a lot of faith in their deckhand and they must work as a team. 'I'm motivated because I know the fellow up on top [his deckhand] wants to go home with a quid. It's all teamwork and my deckie is paid based on how much I catch. Without him being on the ball, I can't perform to my best and vice-versa.'

Because the divers are working in remote areas, they are continually modifying the way they dive to become more efficient. 'I used to jump into the water and bring the ab to the boat,' Glenn says. 'Now I stay on the bottom and the deckie drops a line down to the bottom, I hook the bag and put air into it, then it shoots up to the surface. So he's always following me and my bubbles. This means I've got more "bottom time". I'm even using scuba now, especially for greenlip.'

Glenn lives 200 kilometres from Port Fairy and travels there to go out for one day if the conditions are right. 'When I do work it's very rewarding financially. On a good day I used to get between 350 to 700 kg. Now, on a good day it's maybe 300 kg. I can still get 500 kg at, say, Lady Julia Percy Island but we've only got a certain quota for that area. Unless I put in a lot of hours I'll only get a couple of hundred kilograms.'

While Glenn's dad Lou used to wear a football jumper to ward off sharks, Glenn uses a shark deterrent system. These work by utilising an electric field that induces spasms in the shark's snout and affects its senses. It does not attract sharks to the area or harm the animals or environment. 'For years I didn't use one and I was scared when I was diving at the end of the day, which is dinner time for them. Seals bite my flippers and me all the time and every now and again I'd feel them nip and think "that one felt a bit different". I'd look back just to make sure.'

The only issue with these devices is that they tell the sharks there's something nearby, so if the battery is flat or you turn it off, the shark will come and investigate. 'One day mine filled with water and I thought, no way am I getting back into the water because all the sharks around knew that I was there. I had nothing to repel them with so I thought I'd be lunch.

'I had a recent scare at Julia Bank, which is past Yambuk, 3 kilometres out to sea. I was in 50 feet [15 m] and there was a 9-foot [2.75 m] shark coming towards me at arm's length. I put my knife out and ran it along its belly as it went past to keep it away from me. I did this three times and it got to the stage where I couldn't work because it kept coming back at me. In the end I thought, "Yep, three times — time I was out of here." Funny, I wasn't scared at all. It was a tiger shark and they're man-eaters.'

Glenn also wears a 'pony', which is a small scuba tank with an independent regulator that's turned on and ready to go if his compressor stops. There's also a reserve on the compressor, known as the 'coke can', which is ready to pump air down if the compressor stops.

Clockwise from top left: Glenn Plummer.
Glenn Plummer's tools of trade. Glenn
Plummer's abalone boat ready for work.
Photo source: Glenn Plummer

Glenn recalls one occasion at Bridgewater when he launched his boat and got over the first eight waves. 'The ninth was probably 10 feet [3 m] and knocked us over. My boat ended upside down and bins were scattered everywhere. The other boats were waiting to go out but instead they ended up collecting my gear from the beach. Scott McRae was diving at the Horseshoe a few months later and found my hose reel.'

Another day at Bridgewater, Glenn had been diving for eight hours when his motor wouldn't start. By 5 pm he was worried they'd be stuck there for the night. 'We grabbed the oars and thought we'd let off a flare, but the economist in me looked

at the range of flares and picked the least expensive and let that one off. I didn't think anyone on the beach noticed it.' Glenn then decided to let off a parachute flair — the most powerful flare he had. 'We saw the cop cars coming. Apparently a guy on the beach had seen the first one as he was unloading his abs. He ended up strapping us side on and we went in like a catamaran and landed that way.'

Glenn says, 'It's more dangerous now because you're working deeper and so are more prone to getting bent. There are certain stages to decompress: 20 feet [6 m], 10 feet [3 m] and then surface, but in that time you're not meant to do anything. I don't like "hanging" in 10 feet of water, because I reckon that's like being a piece of burley. I tend to move into 10 feet and work that area to get a few abs. It keeps my mind active rather than just hanging here for 30 to 45 minutes. We come up and suck on the oxygen. I know that's not ideal but it works. I'm told I probably have been bent but I try not to worry about it.'

Glenn is active in WADA, having been treasurer and then president for a year. As president, he represented WADA at a high

level and admits feeling uneasy at times representing millions of dollars worth of licences. He agrees it's a lot easier now with Harry Peeters in the executive officer role and his brother Phillip as president. Glenn also volunteers to do abalone sizing to get data used by WADA.

Like most divers, Glenn rarely eats abalone. 'Every now and again I do eat one that's fried in butter. It needs to be sliced and then hammered, then eaten as soon as it turns white, probably 30 seconds. Any longer and it's a piece of rubber. It's a different taste. Give me King George whiting or flake anytime. The aphrodisiac effect? Well, it must have worked in China ...'

Phil Plummer
Like looking into death

The Plummers are an abalone diving dynasty but, as Phil explains, he wasn't born the son of an abalone diver. That came later. 'The old man only took it up as a full-time career when I was 18,' he says.

Phil was at university at the time and still does a bit of teaching when he's not diving. He works one of his dad's licences and brother Glenn works on the other.

In the early days Phil and Glenn used to argue, as siblings do, but not when they were diving. 'There was plenty of friendly rivalry, especially when Glenn ended up deckhanding for me at one stage. That was a hard period of working together.'

Lou used to take the kids out on a boat when they were children and they would get seasick every time, but Phil didn't get seasick below the water. 'I was better off under the water,' he explains. 'You do get pushed forwards or backwards but that's not the same as going up and down.'

During his diving career Phil has had plenty of close calls, including a number of drowning experiences when something went wrong. One day he was working at Bridgewater in 60 feet (18 m) and came across a small cave. He wriggled into it and was collecting abalone when his deckie gave the signal for him to return to the boat. As he

tried to get out of the cave, his air fitting was pulled off. 'So, I'm in a cave 60 feet underwater, and I had no air.' Phil had to wriggle out of the cave backwards and make a dash for the surface. 'It felt like forever. It seemed to take ages to get out of the cave and then get to the surface without any air. I was turning blue and gagging.' He came up as fast as he could and hit the surface with his body lifting out of the water by at least half. He took a huge lungful of air and scrambled into the boat.

One day at Julia Bank, a seven-gill shark kept harassing him. It would swim away and then come straight back. 'I belted it with my iron bar and it took off and went away, then it turned around and came straight back. I soon got sick of that, so I gave it a really good thump under its throat and it took off. I couldn't see him so I thought, "Right, time to go up." Then I looked underneath me and he's still following. That's what they do. They come up underneath their prey and attack them.' Phil got to the surface and the shark swam away.

Phil's white pointer experience was at Bridgewater in front of the seal caves. He was in 30 feet (9 m) of water with very little weed for him to hide behind. He suddenly felt strange and looked up to see a 15-foot (4.5-m) white pointer gliding slowly past, not moving a muscle. 'It was gliding through the water like a submarine. The only thing that moved was its cold black eye that never left me. It was like looking into death.' At that stage Phil thought the smart thing to do was to get back to the boat, which seemed to take forever. After he calmed down, he had his lunch and went back to work. 'But I was looking over my shoulder the whole time,' he says.

Although the relationship between divers is good — to a point — Phil reckons that point is when you're on a good patch. 'It's like a dog with a bone. You don't try and take its bone. It's a bit frustrating when you find some other diver near you on your patch. After all, it's a big ocean out there. You scratch your head and wonder why they need to be on top of you, but it happens. I know a couple of guys who do that sort of thing. They have a "me, me" mentality. They can be great divers and get good catches, but if they're not going so well on any particular day they're on top of you, trying to harvest your spot.' Phil gets annoyed when that happens because he tries

to leave people to their patches and won't bother them. 'I tell myself that if they have a good day, I should go in and play their game and keep annoying the crap out of them — but I've yet to do that.'

Phil has seen poachers near where he lives. One day he was taking his dog for a walk along the beach and saw a group of men sitting on rocks not far from his house. They didn't look suspicious but his dog charged at them and came back with abalone in his mouth. As the area is a 'no-taking-of-abalone' zone, he phoned Fisheries and kept on walking. Fisheries came down and the group took off, but when they came back to get their abalone they got caught. They were caught with about 30 undersized abalone. 'They'd been here a few days, so it wasn't their first attempt and it was a fairly serious amount. They weren't snorkelling but they were fossicking around the rocks and took absolutely everything. There was more shell than meat in what they took so they're getting nothing out of it, just crippling the industry.'

Phil has been diving for about two decades and is happy to stay in the industry for the time being, although it does depend on what sort of a day he's had. He hopes the fortunes of the industry will eventually turn around but admits it has been a long slow process. 'It used to be a hell of a lot better than it is now,' he explains. 'These are difficult times.'

He wouldn't discourage his children from diving, and says that if they're that way inclined they'll find it themselves. His son enjoys diving and they go crayfishing together, but he doesn't chase him to do it.

Phil says he eats abalone if someone knows how to cook it. He's had some sensational abalone dishes, usually at functions he goes to. 'I'm always in awe of how they do it. They make it look so simple and it tastes amazing.' He thinks raw abalone has an interesting taste, similar to cooked abalone. 'You wouldn't know the difference. Just slice it raw.' Phil recalls that Ricky Harris always ate it raw when he was working. 'He'd sit in his boat eating his catch. I've tried it at the abalone farm, admittedly they were small ones, but after tasting them raw I couldn't understand why you'd bother to cook them.'

8. The deckhands

Garry Bartle

We should have been killed

Garry Bartle is Rod Crowther's brother-in-law and was his deckhand for 14 years from 1986 to 2000. He also did occasional work for Phil Sawyer, Rick Harris, Bob Ussher and Jamie Espie, as well as Victor and Ron O'Brien, and any other divers who needed someone for a day's work.

Garry reckons those years were definitely the glory days of abalone diving when there was big money being made by the divers and a good living for the deckhands. If the divers caught their 20-tonne quota, then the deckie got, say, $1 per kg, or $20,000 for the quota. However, Garry maintains this is not a lot when you think about the responsibility deckhands carry by having the diver's life in their hands. This is especially the case when

the divers could make perhaps $1 million for working maybe 40 days a year. 'When people talk to you about being a deckhand and you say you make $1000 a day, they think that's fantastic. But you might not make that $1000 for another four or five weeks.'

The flexibility of being a deckhand worked well for Garry, who also had a gardening business and caravan yard and could go out when conditions were right for diving. 'I was pretty well available any time. I also had a bit of common sense, which you need. Some of the deckies were a bit younger and they might have been out on the grog when they got the phone call in the morning, so they didn't want to go out.'

Garry won a dive course in a raffle, but that did not convince him to take up diving. 'I'm a committed fisherman. There's a lot of fish and sharks down there, I'd rather catch them. I'd also rather be on top of the water than underneath it — as much as I'd love to get down and have a look. I've been down to about 30 feet [9 m] and it was fantastic but diving was something I didn't want to start. It meant more money, more energy. I thought about it but decided it just wasn't going to be me.'

Admitting he likes a challenge, Garry believes the sea is always a challenge. 'The first or second day I went out with Rod, we were at Bridgewater. I was a bit green and didn't know much about the swell or wind direction. We came inside Bridgewater and I looked out and saw Phil Sawyer and he couldn't hold anchor and was drifting out to sea. I gave Rod's hose a tug to tell him what was going on. But of course we were on a patch of abs and it was dead flat where we were — away from the swells.

'Rod came back up half an hour later. We were in our protected area, but the seas outside were about 20 feet [6 m] high. They were dumping like big grey monsters and we were 20 miles [32 km] from Portland. How we got back I just don't know. We should have been killed. At one stage we were coming down a wave and the boat did a 360 degree

and then popped back up again. At that stage I thought we were goners and wondered what I'd got myself into.

'When we got back home, Rod's wife said to me, "What did you do, have an argument with him?" Apparently Rod was sitting in his chair in a daze, like he was thinking how lucky we'd been.'

Another time they were just past Lawrence Rock coming into Portland. Garry looked out the back of the boat and saw what looked like a pool of oil. Rod thought they'd probably 'done' a motor but it wasn't oil on the water, it was blood from a 7-foot (2.13-m) sunfish they'd run over.

Garry was always aware of his responsibilities as a deckhand. 'My worst nightmare was pulling the diver back up and only half of him being there. It was always in the back of my mind. Or the diver might get stuck and I wouldn't know.' He used to worry about these types of scenarios a lot, and says there needs to be strong communications between the diver and deckhand. 'These days they have a virtual reality set-up so the diver and deckie can talk. Although these were available when I was there, none of the divers had them. I couldn't understand why not.'

Garry remembers being at Whites at the back of the boat cleaning when he looked across to see a 50-foot (15.24-m) crayboat up against the rocks. 'This big whale had rolled up against the side of the boat and it probably thought it was the side of a cliff. The fin was maybe 5 feet [1.5 m] from the deckie's head. I don't know what he said, but the look on his face said it all.'

The era when Garry was decking was towards the end of the old style of diving. By the time he left there was a new breed of diver, often someone who leased a licence from the owner. 'At the time the owners were making a lot of money but these divers came along and leased licences. They were fast divers.'

Garry says people would comment that one person or another was a bit funny, but to him the men were normal: 'Just hardworking blokes who were in the right place at the right time. They just went out and did it. Sometimes I'd go out, look around and think, "Geez, I wouldn't want to jump in there" — but they had to. They had no choice. The money was always a big incentive.'

Although Garry hasn't been in the industry for quite a few years, he reckons there must be patches where there is still a lot of abalone. 'If they're picking up 120 kg in 40 minutes, then there must be lots of them there.'

Roy Clapp
Canoodling, wobbegongs and Drain Bay

Roy Clapp was Andrew Coffey's deckhand for around 10 years in the 1980s. Roy's parents originally owned the land where the Southern Ocean Mariculture abalone farm is at Port Fairy. The land is called Drain Bay because it has a drain going through it from Goose Lagoon. The family still has around a hectare of beachfront land near the farm that takes in Blue Nose, the back of Long Quarry and the start of Crags Bay.

Roy's family knew the Coffey family and one day he called in to see his parents and Andrew was there. Andrew asked him if he was looking for work. 'I said I'd give it a go. Had I known then it was one of the most dangerous jobs around maybe I wouldn't have taken it on.'

The work was hard, says Roy. You had to be mentally alert and watch the changing sea and weather all the time. 'We used to do a lot off the beaches and rocks. We'd cart the bins and compressors down by motorbike, which was hard work but fun.' Andrew always had a dog with him, a Dalmatian. 'Some days it was annoying, but it was okay.'

Roy starting working in the industry just after they weren't allowed to shuck the abalone any more and before limits came in. He recalls catches of anywhere from 100 kg to a tonne. 'In those days we were allowed to go out and get as many as we liked. We had open slather.' Fisheries used to patrol the waters in helicopters and Roy reckons they'd always approach up wind so the people on the boats didn't hear them. 'Then they'd buzz us and scare the living daylights out of us.'

Roy remembers the good times, like playing with baby seals at Lady Julia Percy Island. They were like puppies and would come up onto the deck at the stern and allow themselves to be patted.

In an industry dominated by men, occasionally women came out on the boats while the men dived. Roy recalls one day when he and Andrew were at Bridgewater, along with Phil and Lou Plummer. Phil had his girlfriend with him and they drew their boats together for a chat while the others dived. At one point Andrew surfaced and asked what was going on: 'Some canoodling?' Then he went back down and picked out a 3- or 4-foot (1-m) wobbegong shark (a harmless bottom-dwelling shark) and threw it into the Plummer's boat. 'There were arms and legs going everywhere. Phil was laughing because he knew what it was. His girlfriend was screaming. I'm cacking myself laughing. She wouldn't come out with us after that. It was the funniest thing I've ever seen.'

Roy believes that if you treat the sea right, it will treat you okay. Sometimes they'd put the boat into a spot and Andrew would go down diving. 'I'd move the boat and anchor up and find where the good spots were to park. Andrew would come up and say, "How did you get the boat in here?" I'd say, "You go back and get the abalone".'

While there were lighter moments, there was also lots of 'hairy' times. Roy remembers that he'd be sitting in the boat when suddenly a freak wave would break over the top, filling the vessel with water. 'The old heart pounds for a while, then you get the hell out of there. At those times we'd be glad just to get back to the wharf.' One day their motor broke down in rough seas and they didn't think they'd get back from Lady Julia Percy Island.

'We should've stayed out at the Island but we decided to come home. While we couldn't get onto the Island, we always carried more supplies than we needed — just in case.'

Once when he, Andrew Coffey, Lou Plummer and Phil Plummer were at Blue Nose, a hailstorm with hail stones as big as golf balls hit. They didn't know if they were going to make it back safely to Port Fairy. 'You get through it and then all of a sudden you're thinking about going out the next day.'

Roy used to do some crayfishing with Phil Plummer because they were both qualified coxswains. 'At one stage I thought I'd get my own licence. Andrew had taught me how to dive properly, even though I couldn't actually dive for ab. He treated me like a son, and it was all good.' But Roy decided not to get a licence. He suffered badly from seasickness and started taking medication to combat it, which worked for a while but then he started getting it again. When he found out he had ulcers, he gave it away.

Phil Haywood
Lucky Phil

Phil got into the abalone industry through his mate David Forbes. The men grew up together and did a lot of diving. Phil worked as David's deckhand for five seasons and then for Bob Ussher. Phil is a quiet, unassuming man who prefers to let others have the limelight. He gives the impression that he does his job well on the day.

The main reason he was attracted to abalone was the free and easy lifestyle. He also liked the fact that no two days were the same. But according to Phil, the job can be stressful because of the responsibility of being in charge of the boat.

Phil says some of the catches with David were the best on record in the Zone. He remembers one big day when the boat held so much abalone it was only just out of the water when they got back. 'I was packing ab all the way to shore and for an hour after we got back. I had a trail from the packing board at the back to the boat controls — it was like a goat track through the abs. They were spread out on deck and everywhere.' Divers like Bob Ussher would catch a tonne of shucked meat — a massive amount. 'These guys would do this in the morning and go to the pub in the afternoon, then wonder why they got bent. Often they'd only go just out off Portland — not far at all.'

Because David had a system of using small bags, life above the water was always pretty busy, even when Andrew Beauglehole was on board as the second deckie and driving the boat. Small bags meant David was doing a bag change every four to five minutes or so and, although for David it was an easier way to work, it was hectic on board for Phil. Because they worked with small bags and had more frequent change-overs, there was always a bigger risk of something 'stuffing up'. As he says, he always had to be on the ball and they never anchored and worked live. It was 'busy, busy, busy'.

'One day at Crags we got caught in a bombie [bombora] and I had to back through about 2 metres of white water while Forbsie was on the bottom. We ended up with a metre of water in the back of the boat and I was at the front trying to control things. I had water coming inside my gumboots — there was a bloody lot of water on board. Another day we got caught inside with Forbsie on board the boat when we were having a look at a spot just off Nelson. We got the windscreen smashed when a wave came over the top.'

Because David doesn't like to work deep, it was trickier for Phil as deckhand because there is more risk of waves and reefs. Phil thinks it's a clever way to work. 'I had to shout him a few beers after running over the air hose because he had to buy a new hose and that's expensive. But this happens when you work live and in tight spots.'

Phil also worked for Bob Ussher. 'Bob was a good bloke to work for, but a pretty unusual character. But then all ab divers seem to be characters. Bob's done it all, but he's paid the price for it now.'

Phil left the industry around 2005.

Kim Heaver
A friendly rivalry

There's something about deckhands being quiet, unassuming men, and Kim Heaver is no exception. Kim was Ron O'Brien's deckhand for 15 years, from around 1970 to about 1985. He used to dive after school in the mid 1960s when there was no such thing as an abalone industry. For him, it was just what kids did back then.

While he was at university he did a lot of diving with a group of students. Then they found out that licences had been restricted and no licences were available. Kim wasn't worried that he couldn't buy one: 'If I could have, I would, but I wasn't worried. A lot of blokes jumped up and down and weren't happy, but it didn't faze me.'

After this he was at a loose end for a while, before Ron asked him if he wanted to work as his deckhand. The lifestyle suited Kim just fine, working only around 50 days a year, plus he and Ron were good mates. Kim worked on both a set wage and a percentage of Ron's catch. 'When I first started, a lot of the time we couldn't even sell the catch because there wasn't an established market. We'd sell it to local Chinese restaurants. Buyers would come to the boat ramp to meet us and want to buy it for cash. A lot of the time when we couldn't sell it, we'd throw it back in the sea.'

As a deckhand Kim had responsibility for everything in the boat, while Ron was in charge of getting the abalone. He would have to move the boat around to different spots and make sure everything was running smoothly, including keeping the compressor going.

The waters off Victoria's southwest coast are notoriously treacherous and Kim remembers dangerous incidents happening all the time. 'The weather was unpredictable.

We'd go out to Whites Beach and the weather would be nice and calm — a beaut day. Then it would change and we'd have to go home 24 miles [38.5 km] through big seas with a full load of ab. Sometimes it was an ordeal. A few blokes even lost their boats.'

According to Kim, there was always a bit of fun among the divers. 'Blokes would go in the opposite direction to fool you. There was fun and games on the water, but it was always friendly rivalry.'

Kim is now in the 'bare boat' industry renting yachts.

Bob Hope
The provedore

When Bob was a young man he decided to hitchhike around Australia with his girlfriend. They ended up in Portland. He got into the abalone business when he met Phil Sawyer through a mutual friend who was shelling at Portland. 'I got here in the early 1970s when the original abalone business was interesting. There weren't any rules and these guys had so much money that whatever rule there may have been, they were able to buy their way out of it.'

Bob was a sheller working primarily

with Phil Sawyer. He also worked with Ron O'Brien, Bob Ussher and Rick Harris. According to Bob Phil Sawyer 'was a character. He was a science graduate, a philosopher, a boatbuilder — the whole bit. He was a good guy in many respects but also a pain in the arse a lot of times.' According to Bob, Phil was one of the rare breed of diver with a tertiary education. Most were surfers or tradesmen. Most of them were from New South Wales whereas Phil came from a fishing family in Port Lincoln, South Australia. 'He had a nose for fishing even though he did other things as well,' says Bob. 'He built one of the first trawlers. He was a pioneer in that he tried to do lots of things.'

Bob says Phil was an innovator. He was the first person to get a big boat in Portland, buying his brother's 40 footer (12 m), *The Tara*, which was a big fibreglass boat that would never go fast enough. Bob reckons it would have needed another 2000 horsepower to make it to go faster. 'The idea was to be more flexible. Instead of having a speedboat that was rough and you'd get wet and spill your coffee, Phil had a decent-sized boat. We drove around in this 40 footer when everybody else was in 20 footers.'

Konrad Bienssen, who had just retired from the Fisheries Department, came on the scene at the time that licence transferability was introduced. Again Phil pushed the boundary when he figured out there was no law to say you couldn't have two divers on the one boat. Phil bought Tony Jones' licence and Konrad worked the second licence. Bob looked after the two divers and the boat. Once they got established, they found out it was in fact legal. 'Phil just kept ringing people, hassling them, talking to parliamentarians and eventually it was obvious that it didn't really matter.'

Bob remembers they used the first shark cage in Victoria. 'It was pretty good,' Bob says, 'but there's bugger all sharks around here.' So they modified the shark cage and turned it into a pick-up truck. 'It was a hydraulic thing with hot water and air and had the capacity to lift huge amounts. We made these enormous bags. Previously the divers used to work with 30-, 50-, maybe 80-kg bags. Bags of 100 kg were just too hard to manage, but these guys with this modified shark cage had 300-kg bags. The two guys could get 150 kg each without having to surface.'

These days Bob works as a ship's provedore in Portland.

John Jehu
A good bunch of blokes

John saw many divers and deckhands work the Western Zone in his 19 years as a deckhand. He also witnessed a lot of changes in the industry. He decked mainly for Murray Thiele, and he also did stints for other divers as the work came up.

Andrew Coffey introduced John to the industry when they both worked on the *Warrnambool Standard* newspaper. John started out decking for Andrew, and then began working for Murray. 'With Murray living in Avoca, I kept an eye on the sea and weather down here. If it was a good diving day I'd ring Murray and he'd come down, pick me up and we'd go out for the day.'

The irregular nature of the work suited John well, as he was able to combine decking with his night-shift job at the *Standard*. Often his working day would begin at 9 am on the abalone boats. Once he finished around 6 pm, he would have dinner and go to work at the newspaper until anywhere between one and four in the morning. If the weather was good for diving the next day, he'd do it all again. Although the regime was tough, the money he earned as a deckhand was very good. 'Working on the abs set me up. Otherwise I wouldn't be retired now.'

John enjoyed both the lifestyle and the men in the industry. 'They were a good bunch of blokes to work with. You'd get all different types because of the itinerant nature

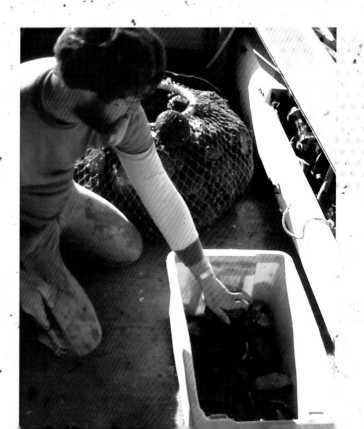

of being a deckhand. The divers would get guys out of the pub for a day's work. One bloke was a fisherman who was one of the toughest guys you'd see. When there was a thunderstorm he was the first to pack up and go home. He'd be hiding under the bed.'

Like any industry, they had their funny times. 'One day we'd been working Lady Julia Percy Island and we were back at Port Fairy throwing the ab bins up on the wharf. A tourist walked past and asked if we'd caught anything. At the time Murray was about 60 years old and I replied to the tourist, "Nah, me mate was too old and we never caught a thing." There was all this ab around and all the divers were falling around killing themselves laughing. The poor tourist didn't have a clue.'

Poaching is an insidious part of the industry, but there were some light moments in their dealings with poachers. 'One of the divers used to cut the valve stems out of their car tyres and then ring Fisheries to come and pick them up.' When mobile phones came in John recalls how Fisheries would phone asking if we'd seen a certain boat go through or what certain boats were doing in the water. 'We had a good relationship with Fisheries.'

One day they were at Goose Lagoon when they saw poachers setting up. 'We told them that they could only get five abs and to move on. They reckoned they couldn't understand English. We packed up a little later on and moved to another bay. They followed us. We went back and they had all these bags of ab hidden in the water with rocks on the bank pointing to where the ab was. I moved all the rocks and stayed there waiting for Fisheries to arrive. The Fisheries guys took me into town and then went back and waited for the poachers to return. They said the poachers spent a heck of a lot of time walking round trying to find the ab. They eventually found the bags and, as they were heading back to their car, Fisheries nabbed them. They thought it was a hell of a joke.'

John says that when he did his training and then wanted to get his coxswain licence, the authorities in Melbourne didn't want to give it to him because of the sporadic nature of decking work. Their instructor told them, 'These blokes know more about the sea than any cray bloke does because they work around the reefs and rough water. They can read the sea far better than anyone.' This must have been enough for the authorities because they were convinced and gave him the licence.

Divers and deckhands must be adaptable and able to act quickly to fix equipment as both their lives and livelihood depend upon it. 'You had to be a mechanic on the water and keep everything running. If something broke you fixed it out there at sea, otherwise it was a lost day.'

John recalls how Murray Thiele suffered from salt inhalation at one stage and this was traced to the mouthpiece he was using, which brought water into the mouthpiece and he 'got crook from inhaling salt'. The local doctor identified the cause and it was solved with a change in mouthpiece. 'It was all a learning experience,' says John.

No one disputes that there was friction between the Portland, Port Fairy and Warrnambool divers and this was exacerbated when a diver worked outside his own turf on a reef in another area. 'Port Fairy divers used to hate the Portland guys coming over and working at, say, The Crags and vice-versa.'

John loves eating abalone. He reckons the best way to eat them is to cut them in half. 'Give them a good pound with a steak mallet and drop them onto a barbeque hot plate.

As soon as the edges curl, turn them over. As soon as you can cut through them with a spatula they're ready. It's like eating into soft cheese. Beautiful.'

Robert Rendell

Eating abalone is like eating your thong

Robert Rendell has been involved in the fishing industry for around 30 years. He loves the ocean and says it's been good to him over the years. Although he loves the ocean, the lengthy periods of time away from his family and home when he was crayfishing and deep sea trawling began to take its toll around 1999. When an offer came up for a deckhand job in the abalone industry, he jumped at it. He says the lifestyle suits him well.

Robert cut his teeth in the abalone industry working for David Land and then Peter Ronald for around six years. For the past six years he's been Peter Riddle's deckhand — and loves it. Although he still does some shallow diving for fun, he's happy working on the water rather than under it and he's not tempted to jump camp and become a diver. 'The shallows would be okay but the deep water — that's different. There's a lot of things down there I don't like.'

In the years Robert has been in the industry, he's seen many changes. 'When I first started we could leave Port Fairy boat ramp and go diving anywhere in the Zone. We'd come back with at least 500 kg every day. Now they tell us where to go and how many we can take from a certain area. There are size limits and we have to scan all our abs.' Robert also has to tag the abalone and complete paperwork, recording the amount of abalone taken from which areas, sizes and reef code information.

Robert reckons Peter Riddle is probably the best diver in the Zone, along with Rob Torelli. He says Peter is a true professional at his job and a great boss. 'I enjoy working with him because he's so professional. I don't really have any scary times with him. All his gear is spot on. If something breaks down or goes wrong, he fixes it straight away. But then both our lives depend on it and you only get one chance. He's good to work for and pays

well. We have a lot of respect for each other.'

Peter and Robert dive on multiple licences and this means going out around 50 days a year. They only fish live these days — no anchoring. This means Robert drives the boat and follows Peter around all day. Thanks to the accuracy and reliability of the weather maps and data available on the internet, they know a week in advance when they'll be going out.

Peter Riddle is well known for being calm and quiet. But Robert remembers how Peter's reputation was put to the test on the first day they went out together. Robert had previous experience as a deckhand and so Peter only gave him a short briefing: 'He said not to worry because I'd get the hang of it pretty quickly. Just before he jumped in, he said, "I'll know when to come up. I do not want to be pulled up for anything. Understand?"'

So Peter dived and Robert went about his job and followed Peter. Not long after, Robert was shaken to see a fin going around in the water. Then he saw another three fins. His first day on the job with Peter and already he sensed danger — sharks. The first thing that crossed Robert's mind was whether to let Peter know, but he remembered he'd been specifically instructed not to pull him up for any reason. But sharks? That's serious stuff. 'When they got closer I realised they were killer whales, which are massive, and could have done anything to him or got tangled in the hoses. I was shitting myself at this stage. My first day and we had killer whales and Peter in the water. Well, I did pull him up and he was pretty darn cross. He said, "I've just got down there, what the hell do you want?" I told him that three killer whales had just gone past but he didn't want to know about it. He went straight back down.'

Robert's job is to look after Peter and the boat. 'It can turn pretty nasty, pretty quickly.' Peter usually only surfaces once a day. He's in the water at around 9 or 9.30 am, comes up for lunch at midday and then dives again until around 4 pm. The depths he dives also contribute to how long he stays underwater, but in general he starts diving deep then finishes in the shallows where he decompresses at the end of the day. Peter takes oxygen on the way home when he's been diving deep to help his recovery.

Abalone divers traditionally have good but stock-standard boats into which they stack the bins of abalone. If they have a good day the boat can be stacked high with bins. But the sea is an unforgiving master and in rough seas the bins stacked in the boat can create a safety hazard. When Robert first started in the industry they used 50-kg bins but now they use smaller 28-kg bins to comply with WorkCover legislation. 'The new bins are much easier to pack on deck. After using the 50-kg bins, we thought we'd have the smaller bins everywhere. But they are light and easy to handle on the boat. They're good.'

Those seven hours each day create a solitary work environment for both men. Robert keeps in touch with Peter by sending a slate down to him with information about how long he's been down and the depth he's diving, although Peter keeps himself informed about this as well through his diving watch. Robert sends down one slate every hour or so. A typical message might be: 'You are at 45 feet [12.5 m], 200 kg of abs. You have been working NE, now heading E.'

Robert is a farmer and says being a deckhand gives him a break from his farm. 'It recharges my batteries.' He believes being a deckhand is an important job. 'When the divers are in the water, they've got enough to think about. They're picking up $10 notes from the bottom and my job is to look after them. It's all about survival.'

Robert's done all right out of being a deckhand, but he doesn't eat abalone. 'Don't like them. May as well eat your thong, I say.'

Herbert Walden
A lot harder than a pick and shovel

Herbert Walden never set out to be a deckhand. In fact, he had a good job as a joiner, but things changed when the company he worked for started a redundancy campaign. 'I thought: "This is no good." So I had a yarn to Lou Plummer about working for him as a deckie and he said he'd take me on. I took a fortnight's long-service leave and I didn't mind it at all. After a fortnight I went back to work and took a redundancy as this meant one of the younger blokes could keep working — the blokes with a mortgage and things like that.'

Herbert met Lou when they were both in a skin diving club and he had been Lou's neighbour since the mid 1960s. Even though he worked for the Plummers, he was never tempted to take up professional diving. 'If a man works on a pick and shovel all day he gets fit and tired. But diving is a lot harder than even that.'

Herbert worked for Lou for two years and then Lou's son Phil for about five years. As a deckhand, he saw his job as making sure things were right for his diver such as that there was no kink in the hose line, watching out for potential hazards, cleaning and packing the abalone and putting them into a bin. Then when the diver surfaced, he had to pull him in and winch the bag on board before giving the diver another bag. It sounds a straightforward job, but according to Herbert it was all about keeping the diver safe and on his good patch.

Herbert reckons Lou was always a hard worker. He remembers him working with Andrew Coffey. 'Whether it was rough or not. If they couldn't get out in their boats, they'd take a compressor down on the rocks and work from there.'

Working with Lou was hard work, but there were some lighter moments. At one stage Lou had an ear problem that affected his balance. 'He'd go down, then come up and go back down again. When he came up, he'd start chundering over the side of the boat, then go back down again for maybe 15 minutes before coming back and chundering again. One time he lost his false teeth over the side and couldn't find them. He was always looking around for a fish wearing his false teeth.'

Herbert recalls how they would average perhaps 600 to 700 kg a day at The Crags and sometimes they'd get their tonne. He worked most of the west coast with the Plummers, from Warrnambool to Port Fairy and up the coast to The Crags and Lady Julia Percy Island, and also at Portland where they would work out of Bridgewater Bay and along the reefs. 'My favourite spot was The Crags. It was close to land and pretty safe, but we needed to keep an eye on the breakers. It was also good fishing.'

Most divers would pick and choose the days to go out, and Herbert says some would go out no matter what the weather was. 'At the end of the day if they had maybe 200 to 300 kg, then it made their day. They were happy.'

He says many of the divers took risks such as diving in 80 feet (24 m) then, instead of moving up into shallow water to decompress, would keep working at that depth. They would come up, go home and then go to bed with an oxygen bottle.

Herbert lives in happy retirement at Port Fairy.

Dan Hoey
These guys are different, all right!
Dan doesn't like being called a 'deckie'. He reckons that because he does so much more than just work the boat he is the skipper for his divers.

Dan has been skippering for abalone divers for around seven years and works with Rob Torelli, Jason Ciavola and Craig Fox. He is a chippie by trade, and after a divorce decided to have a life change and started working on abalone boats to earn his sea hours in order to get his qualification to do charter work. As well as working with the abalone divers, he now runs a boat charter business. He believes that most people haven't got a clue just how hard the work is on board an abalone boat, and that until you get good at the work it's a two-person job.

Dan knows the sea well and has a great admiration for abalone divers. 'They are a special breed. To spend that amount of time on the bottom and hold their nerve takes a special person. To dive in the seal colonies and work with very little visibility … well, the average person wouldn't do that. It's a big risk.'

Dan recalls working with Rob Torelli one day in deep water at the back of Lady Julia Percy Island, which is known for being a white pointer shark location. The day was flat with good visibility; a good one for abalone diving. Dan was doing a bag change when Rob pulled on the hose. Dan clipped a fresh bag onto the line along with a slate so that Rob could say something to him. When the empty bag came up there was a note on the slate saying there was a shark below.

Dan became worried for Rob and realised he would need some sort of defence against the white pointer, so he clipped a boat hook onto the line and sent it down to Rob. In the deep water below, the shark circled Rob six times before the shark decided there wasn't anything to eat and left.

Dan says Rob came up to the surface and they talked about what had happened, but he wasn't phased by the encounter. 'He wasn't pale white or shaking or anything,' according to Dan. The men went around to the other side of the island, where Rob jumped back into the water and fished for the rest of the day.

There was another occasion when conditions were so bad Dan reckons the men shouldn't have been working, because visibility was down to around a metre. 'We both needed to work that day, so we did,' recalls Dan. The men worked out of Portland and went on to Cape Bridgewater. Dan recalls how Peter Riddle went past and Dan and Rob pulled up alongside him for a chat before starting work. The men told Peter they had seen a dead blue whale that was banging against the nearby cliffs and rocks.

Rob went to work at one of the renowned white pointer shark spots, where there was poor visibility and a large swell. According to Dan: 'Rob jumped in and went to work. As he's chipping abs off, there's chunks of whale floating past. It was the biggest burley trail I've ever seen!' With such low visibility and the gunk at the bottom, Rob wouldn't have been able to see a shark. However, despite the conditions and burley trail of whale flesh, the men completed their day's work.

9. The processors

Joe Milani, South Canning Pty Ltd

It was like they were running a marathon

Joe Milani has been around the abalone industry since the 1970s when it started to get serious and transformed from being a cottage industry into a full-blown, sophisticated industry.

Joe is managing director of South Canning Pty Ltd in Portland, the first company dedicated to processing abalone. According to Joe, there were businesses involved in general food manufacture and/or processing that were dealing with the Asian markets, and they found their customers were asking for abalone. They were getting their stocks from other places such as California, South Africa and some Middle Eastern countries but those sources were becoming depleted. At the time Chinese nationals were migrating to Australia and word got out that Australia had abalone.

The company Joe originally worked for had a small abalone canning business they had purchased and then closed. 'The people I was trading with said they still wanted the product and talked to me about it because I was a food technologist. At the time I knew how to put things into a can, but not abalone. Other canneries were doing tuna and lots of other things and processing abalone as a sideline. I remember a well-known fruit processing company that bought a cannery in Collingwood. Within a few years they shifted it to their jam plant on the outskirts of Melbourne but it closed within five years. When the strawberries came into season, they put the abalone in the freezer and you just can't treat abalone like that.'

Joe maintains that wild abalone is a complicated product to process, which is why processors turn to farmed abalone. It's also very sensitive to its environment and experiences environmental changes very quickly. 'It's probably the "canary of the ocean". It becomes stressed quickly and that's why there are ebbs and flows in production.'

In the early days when the divers weren't full time, it was a stop–start operation. Joe says some of the young divers were dysfunctional and into anything and everything — women, good times and even drugs. For these men it was all about lifestyle and earning quick money to enable them to support this lifestyle.

According to Joe, up until 1984 abalone was fetching between 50 cents and $1 a kg, and by 1986 it had reached $2.50 kg. 'That sounds a lot, but it really wasn't. In a day they would earn $100 — about what they earn in other jobs in a week. But it was a lot of work and there was a lot of risk. They had poor equipment and no OH&S.

'It wasn't until the mid 1980s that there was any sophistication in regard to marketing and manufacture. These guys then started getting more organised and it went on from there: $2 to $5 to $7 to $10 kg.' In the early days there was a lot of uncertainty in the industry, such as no licence transferability, and if a diver had issues with his licence the government took it from them. 'Licences that had been worth $100,000 were suddenly worth $1 million. They were still the same men, but lots of money was being thrown at them by an industry that everyone was jumping into.'

Because of the *ad hoc* nature of the industry there was no permanency in the arrangement and divers could suit themselves who they sold stock to. 'We got the product depending on whether we offered the money they wanted and lost it if someone offered more.' These days South Canning gets its product from all over mainland Australia and Tasmania and also buys farmed abalone. 'Wild abalone is unpredictable with its supply and price, and it is a variable product — unlike farmed abalone.'

Being involved from the outset, Joe found he was working with the divers one-on-one to secure supply. He says it was more a money-driven relationship than friendship. 'It was a matter of whoever paid got the raw materials.'

Joe didn't form friendships with many divers, apart from Dick Kelly and Derek Fieguth. 'When Dick came home from the Korean War he became an introvert and didn't mix with many people. I think I was one of the few he trusted. Derek was probably one of the most educated of the divers and he was probably one of the best fishermen. He was an intellectual, which doesn't always define an abalone diver.'

Joe has great admiration for Bob Ussher, who he maintains was 'one of the early divers and probably one of the best'. Bob, he says, was a 'basic sort of bloke' who was always different to the rest of the divers.

While they played cards and sometimes got into mischief, Bob worked. 'Bob's former wife Joyce would ring up and ask if we were buying abs today. If we said yes, she would tell Bob to go to work.'

There is no denying drugs were around the industry in the early days. As Joe says, 'You've got to remember that time and money are a dangerous combination when it's unfettered.' He recalls how the divers used to have meetings at their homes. 'I can't stand smoke, let alone dope, so if someone was smoking I had to sit in the door and talk with them. How they dived and survived, I just don't know. One of them would drink a bottle of vodka while they were having a meeting and he'd be rolling a joint at the same time as smoking one. The good thing was that they were harmless.'

When abalone started to be worth something the divers became very coveted people. They were given everything. There were no rules in play to maintain supply so all sorts of deals were done. All of a sudden these guys became rock stars. This is around the mid 1980s when ab was worth $2 to $3 kg. Some of these guys were earning $150,000 to $200,000 for 40 days' work, which was a lot of money back then.'

These were indeed heady times and the abalone cash-splash had its consequences. In the 1990s there was a big tax investigation and one Central Zone diver even had a spell in jail but was let off because he admitted to his foolishness.

When so much money is involved, invariably there will be jealousies, underhand deals and sniping across all levels of the industry. Joe remembers that, 'One of the buyers was running two sets of books. They had codes, but it wasn't rocket science. Some divers had arrangements with operators that were akin to a parallel trade between deckies, divers and these operators. The product might end up being sold through legal channels or it might be sold through illegal channels. I didn't get involved because I come from a sophisticated industry and I knew it was only a matter of time before they would get caught. I knew our books had to stack up and I know we lost a lot of supply because we wouldn't buy through these channels.'

At one stage there were nearly 200 divers in Portland and Joe says it was like a gold-rush mentality with arguments, fighting and clans. Third parties and families financed many of the divers so many influences came into the mix. In those early days there were no quotas. 'Bob Ussher caught 70 tonnes one year. Derek Fieguth caught 50 or 60 tonnes and when quotas came in he used to finish his in 19 days. Bob worked twice as many days but Derek would only pick the very good days. Derek was the best diver in the district and Bob was the hardest worker.'

Although there was a big disparity between the diving abilities of divers across the Zone, when quotas came in everyone became equal. 'It brought people like Bob catching 70 tonnes down to 20 tonnes, and guys who were catching 14 tonnes came up to 20 tonnes. These guys had been unfettered for 15 years and then all of a sudden they were regulated. What they did was put all their catch data into a hat and came up with the average of 20 tonnes. The big catches of three divers — Derek, Bob and Phil Sawyer — boosted the average. If those guys hadn't been catching four times the rest, the quota would have been 10 tonnes.'

It's well known that these men dived deep. Joe recalls how sometimes they'd come up with their noses bleeding. When people like Frank Ziegler got them using mixed gases and new technology, they could dive in more shallow seas and abide by the rules. Some of the men ignored warnings and dived to their own rules, but Joe reckons many paid the price in the long term.

'They're wrecked now. Some look okay now but that's because when they were diving it was like they were running a marathon — and they were. But they have a lot of hidden issues, like necrosis.' There was an unwritten rule among divers that they wouldn't dive within one hose length of each other and Joe remembers how there would often be a race to come home and get to the supplier first. 'There were clans and they all had their idiosyncrasies. Derek Fieguth, Phil Sawyer and Tony Jones would sit around playing cards all day. Bob Ussher worked at making investments. Ron O'Brien would go and meditate. Dick Kelly would play with his helicopters. Rick Harris would move sandhills around his farm with a big tractor.

'I could earn a lot more outside this industry, but it's an industry I love and I've put so much into it. But for the investment and the amount of work you put into it, there are very small rewards.'

Sou'west Seafoods Pty Ltd

Sou'west Seafoods was created out of the need identified by a group of the first divers and it became integral to the processing and exporting of the Zone's abalone. The business ceased operations in 2013.

When abalone divers arrived in the area in the early to mid 1960s there were few rules. Some had dived for abalone in New South Wales, particularly in Eden, then moved south to Mallacoota, before settling along the Victorian coast around Marlo and Phillip Island. They then moved onto the southwest coast.

2 — The Standard, Saturday, October 30, 1982

EXPORT PLANT OPENS

Port Fairy fishing co-operative Sou'West Seafoods yesterday moved another step toward breaking into the lucrative seafood export trade.

This company, which processes all Port Fairy's abalone and "mice" of the crayfish catch, officially opened its export processing yesterday.

The $150,000 extensions will allow Sou'West to employ two more shop assistants and another three processing staff and about five extra workers during the peak processing season.

Port Fairy Borough mayor Cr Margaret Whitehead who unveiled a plaque commemorating the opening said the co-operative had "shown it was prepared to take the initiative to expand" at a time when "factories are closing and workers are being retrenched".

one that was "truly local" because it had been designed and built by local labor and was processing locally caught products.

Sou'West chairman of directors Mr Len McCall said yesterday about 60 people in Port Fairy were now actively engaged in the fishing industry and related processing. Fishing and its spin-off

business" for Port Fairy, he said. Sou'West had recently been granted a Department of Primary Industry export licence and is working on creating new markets for abalone exports.

Mr McCall said the co-operative would concentrate on exporting whole frozen abalone in the shell

Department will help with markets

Some had diving experience, others were attracted to what they saw as quick and easy money to support their lifestyle. The industry was in its infancy with primitive equipment and the men had little knowledge about marketing their catch. Many were content to work when they needed money and sell their catch as best they could. Other divers had vision.

The Western District Divers Co-operative, 1972–80

A small entrepreneurial group realised the untapped potential of abalone, but in order to achieve its full potential they knew they needed to secure product supplies and establish stable markets. At the same time the divers realised the need for responsible stewardship of the abalone beds in the Zone to ensure ongoing availability of their resource.

It made sense that if they applied the principles of collective bargaining and economies of scale, they would streamline their production costs and maximise the prices they achieved for their catch. They envisaged this operational strategy would focus on pooling their own catches as well as buying product from other divers in the Zone and that this would act as an incentive for divers to remain loyal.

The seeds for the Co-operative were sown.

Sandra Downes, chair of the Sou'west Seafood board, was married to diver and licence owner Len McCall. Before the Co-op was established, Sandra says, 'You couldn't

get the best price for our abalone here in Port Fairy, so we'd have to drive to Melbourne and then drive back. You can imagine how in summer the abs would go off because there was no refrigeration. So Len got the other divers together and organised a co-operative.'

The foundation members of the Co-operative were: Murray and Esme Thiele, Andrew and Valmai Coffey, Dick and Natalie Cullenward, John and Maureen O'Meara, Red Quarrell and his girlfriend, who became his wife, and Len and Sandra McCall. The Co-operative started modestly in 1972 with all the members and their partners pitching in to ensure its success. As the Co-op's operations developed, it became clear they needed premises to operate from and somewhere to store their catches. The divers also suspected they were not getting the best deal when selling their product to processors.

Fishbrooks, which bought crayfish from local fishermen, was operating a depot on the wharf at Port Fairy. The company wanted to get out of the industry and the Co-op's divers needed premises. The wharf building was ideal for their purposes.

Sou'west Seafoods Co-operative Ltd, 1980

When the Co-op took over the lease on the wharf premises around 1980 John Sproal, who had been working at Fishbrooks, was appointed as manager. As John recalls, 'The divers did the diving and I did the rest, which included finding export markets. To export we needed a good supply of product and, while our divers gave us a good supply of product from the word go, I also bought abs from other divers to boost supply.'

Around the same time, in May 1980, the Western District Divers Co-operative became Sou'west Seafoods Co-operative Limited, with its primary objective being to process, export and market frozen blacklip abalone.

Initially the Co-op only traded in abalone caught by Port Fairy divers, but they soon realised they needed more product to grow the business. More product meant more strength and bargaining power. They knew there was always a willing buyer for their catch — it was just a matter of under what conditions and at what price. So they approached divers from nearby Portland who were happy to buy in. Len McCall reflects that there was a philosophical difference between the way the divers from Portland and Port Fairy went about their business in those early days. He believes the Port Fairy divers recognised the need to work as a collective in order for the business to prosper, whereas the Portland divers did not understand this.

Once they were established at the wharf, things changed for the divers. Rather than load the abalone onto a truck and drive to Melbourne at the end of their day's diving, the divers motored into the Moyne River then moored beside the wharf. They would throw their bins of abalone onto the wharf at the back door of the factory. A truck came to the wharf each day to collect the catch and the abalone was sold from there. This made life easier for the divers as they didn't have to travel to Melbourne to deliver their catch, while the product remained of better quality.

The divers demanded to be paid in cash for their catch. This meant the Co-op had to liaise with the bank to ensure the bank had sufficient cash, perhaps tens of thousands of dollars, to pay them. An armoured van would transport the money to the town.

Developing export markets, 1982

By 1982 the Co-op was processing a small amount of frozen product. 'We parboiled abalone in the shell, and then froze it,' John Sproal says. 'We also did unbled abalone meat that was packed in 10-kg cartons,' and that year they secured their first export licence. He worked for several years on getting their Port Fairy premises registered for export with the Department of Primary Industries (DPI). He says the registration process was lengthy and complex because of the high DPI standard and this meant practically rebuilding the premises and fitting them out with new freezers and equipment.

The Co-op focused on processing, exporting and marketing frozen blacklip abalone, primarily to Japan. They sold most of their abalone meat to South Pacific Canneries (Safcol) and Russell Crayfish. They also handled crayfish for a time but found this unprofitable. In those days the Japanese market took whatever the divers caught, regardless of the quality of the product.

John recalls that when they first began processing abalone, their two biggest buyers, Safcol and SPC, were of the opinion that the Co-op would fail. 'They thought our aim to process abalone was pie-in-the-sky and beyond the scope of possibility for a small group of divers and one employee. We proved them wrong.'

The involvement of women

During the early years, the wives became an important part of the Co-op team. They opened a shop at the wharf selling seafood products using byproducts from the catch. Valmai Coffey did the Co-op's bookwork. The women were never short of ideas or enthusiasm and had a reputation for running the wharf shop with great aplomb. They were renowned for their seafood chowder and crayfish rolls, which were especially popular during the summer months when the shop was at its busiest.

They also made a pâte with more of the byproducts. Sandra McCall remembers one time when they tried making crayfish pâte and ended up with a batch of crayfish that wasn't cooked properly. 'We had to bring in a food technologist to tell us what we could do with all the crayfish without throwing any away. They said that as long as we did something with it, like cook it, it would be fine. So we shelled all this crayfish and took the flesh out. We had a big copper boiling up on the stove and made crayfish chowder. We also made a pâte, but it was all just a little side business. We tried to make money out of it, but in the end it wasn't worth doing because there was no market for it among the locals.'

For a time the wives also ran a retail shop in Warrnambool where they sold blue grenadier and a range of other fish.

Sou'west Seafoods Pty Ltd, 1987

It became clear to the divers that their industry was changing long term and they needed to keep up with the times. Achieving and maintaining success in the fishing industry has historically been fraught with danger as product development and reinvestment is difficult. Sou'west was no different. The Co-op realised they could not handle every aspect of the business and they needed to seek expert advice and pay for services they could not do themselves. They needed directors, rather than workers, who could drive policy and the strategic direction of the company, rather than being locked into the hands-on work.

In 1987 they established Sou'west Seafoods Pty Ltd, with the members being the original four divers — Len McCall, Andrew Coffey, Dick Cullenward and Murray Theile; plus Rod Crowther, Rick Harris, Ron O'Brien, Lou Plummer, Jurgen Braun, Clark Smock and Wayne Finley (from South Australia).

Expansion to Awabi Court, 1990

By the late 1980s the business had expanded to the point where it had outgrown the wharf premises. Although they would have liked to renovate, the local council wanted the area to be developed as a tourist precinct and didn't think it appropriate to have an abalone factory on the wharf. They were also worried that development of a factory at the wharf would shift the centre of town towards the river. The original factory building on the wharf is still there, but it's now a restaurant/take-away food business — with no abalone on the menu.

John Sproal was part of the push to move to larger, purpose-built premises to meet the needs of the growing export market. The difficult search began for a suitable industrial property and the present site in Awabi Court, Port Fairy, was purchased in 1990. The company built the current office and processing plant and by 1994 were processing frozen product at their new premises for export to Japan.

The Awabi Court premises allowed the business to expand and develop new processing methods, but there were obstacles to overcome — albeit unusual ones. According to John Sproal, the company had avoided buying a small block because of potential noise and other problems with neighbours. However, the factory is situated next to the Port Fairy cemetery and after rain there were complaints about stormwater

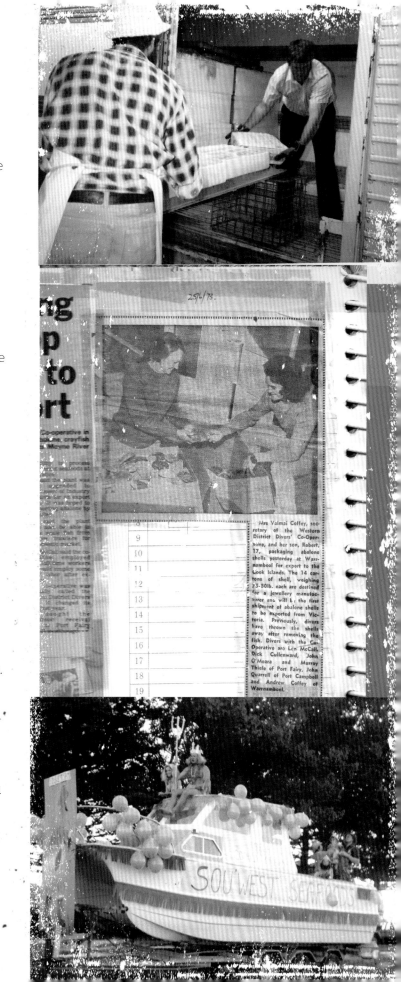

Mrs Valmai Coffey, secretary of the Western District Divers' Co-operative, and her son, Robert, 17, packaging abalone shells yesterday at Warrnambool for export to the Cook Islands. The 14 cartons of shell, weighing 23-30lb. each are destined for a jewellery manufacturer and will be the first shipment of abalone shells to be exported from Victoria. Previously, divers have thrown the shells away after removing the fish. Divers with the Co-Operative are Len McCall, Dick Cullenward, John O'Meara and Murray Thiele of Port Fairy, John Quarrell of Port Campbell and Andrew Coffey of Warrnambool.

from the factory raising the level of water at the cemetery and 'popping' coffins out of the ground.

Sou'west and the divers

Sou'west relied on its goodwill with divers and licence owners to ensure its ongoing supply of product, and has done so since the original Co-op days. Before quotas were introduced in 1988, divers could work anything up to 200 days per year and pull in 30 to 40 tonnes per diver. This made it difficult for the factory to deal with the big peaks and troughs in production.

John Sproal's job was an ongoing challenge, particularly as he had no previous experience in the abalone industry except for a short time as a deckhand. But over 10 years he got to know the Western Zone divers well — warts and all. He recalls how a lot of them had other jobs before getting into the industry, but after several years of diving they found themselves earning such a good living there was no need to work outside the industry. 'The biggest reason was the lifestyle of working 30, 40 or 50 days a year, because they didn't need to work any more than this in those pre-quota days. They had mind boggling incomes — in the millions.'

John remembers the tricks he used to ensure he had sufficient product to meet orders, especially in bad weather. It was well known in the pre-quota days that if one diver went to work, the others went out too because they didn't want to miss out. 'I used to capitalise on this when I needed quota. I'd phone one diver up and say that another diver had gone to work, so he'd go rushing off to get his deckie and boat. I'd then phone someone else and tell him that the other diver had gone out. Before long I'd have the whole lot of them out working.'

He was annoyed when some directors and divers would insist on phoning him at home late in the evening about matters that could have been dealt with during working hours. 'I remember one night it was around 11 pm and I was in bed when the phone rang. The person wanted to know something at that late hour, but I bit my lip and said I'd find out and get back to him. I already knew the answer, so I set my alarm for 3.30 am the next morning and then phoned him. It didn't go down well at all. He didn't do it again.'

When Gary Kenyon first began at Sou'west, it was all about lifestyle for the divers. 'They didn't really want to work for a living, so they used to dive for abs.' Gary recalls how they'd work for a week, then party for a week until all the money had gone, then go back to diving.

Gary had to establish a relationship with the divers as part of his job, but this was sometimes difficult. When he arrived in Port Fairy he started socialising with them, but found this didn't work when he had to deal with them as part of his job. 'We'd have our run-ins from time to time, but on the whole I got along well with them.' Most of the run-ins with the divers related to their catch. Some used to bend the rules. 'They'd try anything. They put lead weights in the bin so it weighed heavier, or they would water down the ab as they were coming up the river — then blame the deckie. Or it might be that the deckie had spilled petrol on a bin and they'd try to put it through, but we'd reject it.'

'The regulations back then were more lax than today and we could get away with more. Those were good times and the staff partied pretty hard. It was nothing to do an 18-hour day then go to the pub until 2 or 3 am, then back at work at 7 am. But not today.'

Len McCall sums up what characterises divers: 'All abalone divers are strong-minded individuals. Extroverts, like a theatre group. All seek the spotlight.'

10. The women

A good life

The Western Zone women are a remarkable bunch. If their husbands are defined as legends and pioneers, then the women must also be acknowledged for the extraordinary part they played in the industry. Their contribution was more than just looking after the family, although of course they did that as well. Many of the women were instrumental in the development of the fledgling abalone industry.

Women like Sandra (McCall) Downes and Valmai Coffey lent their considerable expertise to setting up what was to become Sou'west Seafoods. Sandra was chair of Sou'west Seafoods and for many years Valmai Coffey worked as the company secretary/accountant.

In the early days of the Co-operative, the wives realised there was a potential market for abalone and crayfish byproducts. Esme Thiele, Valmai Coffey, Natalie Cullenward and Sandra McCall were all director/shareholders of the Co-op. They began making crayfish chowder, rolls and pâte and selling it from the shop situated in the old premises on the Port Fairy wharf.

Valmai's expertise was invaluable to the Co-op, but when the directors went to Japan in 1987 to attend a consortium of business people and buyers, the Japanese refused to talk to her. They believed no woman should talk business with them. Valmai was sent away to go shopping with the other women. She was annoyed at the time because she was the only one of the management team who knew the workings of the business. 'Anyway

they muddled through,' she explains.

Natalie (Cullenward) Trengrove, who was heavily involved in the shop, recalls, 'We had all this cray just going to waste so we started using the meat from the head and legs to make chowder. Then we started making crayfish rolls with the meat so that we didn't waste all the beautiful sweet meat. Each woman brought her own skills to the venture. Sandra, Esme and I would serve while Valmai was good with the bookwork. People would come and buy chowder and then sit on the wharf and eat it with a roll.'

Esme Thiele would drive three hours from Avoca to Port Fairy to work in the shop. She remembers how excited they were on the day the shop first opened and they took $1000. Esme didn't do much of the cooking. 'To be honest, I didn't like the smell,' she says. But she made the crayfish rolls and served in the shop. She recalls one day they had a busload of elderly people come into the shop and they all wanted a crayfish but it had to be the same size. The women had their work cut out trying to find 20 similarly sized crayfish, but eventually they were able to send everyone away happy.

Natalie says that everyone had to pitch in when the Co-op first started and that it was built on trust, plus a lot of give and take. 'We would shuck the abalone and when we first started it was about 35 cents per lb [0.450 kg], so you can imagine how many we'd get for 35 cents.' The wives also used to take it in turns to drive the abalone to Melbourne before the Co-op could afford to pay a driver. 'We had to do it ourselves because there wasn't enough money in it in the very early days.'

Sometimes the women went to work on the boats as substitute deckhands. Sandra recalls going out on the boat with Len only as a last resort, when he couldn't get anyone else. She says, 'Len would yell at me, and I'd get seasick.'

There was a brief time when Shelley Ronald was a deckhand for her husband Peter. The Ronalds think she was probably the first female deckhand in the Zone. But Shelley found the stress of trying to keep her husband alive to be too much: she had to make sure the compressors were running well, keep track of the diver, make sure the air hose was clear, know how to operate the winch and the motor. But for Shelley, the worst part was having to pee over the side with a young male deckhand there.

But it was the waiting and worrying about their men that had the most impact on the women. The men didn't always understand what the wives went through each time they dived. They waited for a phone call or for the truck to pull into the driveway.

Valmai Coffey says she worried about Andrew each time he went to work, like all women worry when their husbands go off to do dangerous jobs. She recalls one incident when Andrew was very late getting home and she hadn't heard from him since early in the morning. In those days there were no mobile or satellite phones, so at 10 pm she rang one of the local farmers whose property was near the area Andrew was working. The farmer went out on his tractor looking for him and found that Andrew had become bogged in the sand — but was safe.

Then there was the time Andrew was due to go to dinner with Valmai and friends in Warrnambool. The usual routine was that as soon as they got the boat out of the water at the end of the day's diving, Andrew would phone Valmai and say, 'Boat's on the trailer, I'm on my way home.' Andrew says he must have got carried away with a big haul and it was late by the time he got the boat out of the water. He had to be at the restaurant at 7 pm and knew he should be home by 6 pm. It was nearly 7 o'clock when he rang Valmai and he was still on the boat ramp. She knew what he was like and asked which boat ramp. 'Portland,' was the reply. 'That's 75 miles [120 km] away. We're supposed to be at the restaurant now.' Andrew arrived as coffee was being served. Valmai was not happy.

It was commonplace for the women to phone the other wives asking if they'd heard or seen their respective husbands. Fortunately for Shelley Ronald, Peter had procedures in place and always told her exactly where he was diving and if he moved from one site to the next. Peter also had a documented emergency response plan.

Bob Ussher admits it was hard on the wives and they had plenty of sleepless nights. 'The time I got an embolism and was airlifted to Adelaide, Joyce had to drive all night from Portland to Adelaide. She did it in about six and a half hours.' Then there was the time Joyce was told Bob had drowned when he was working with his deckhand at the Island. He had finished for the day and was about half a mile (800 m) off shore when the motor seized. They paddled the boat back to the Island and, because a storm was brewing, decided to stay the night. They went to the beach at daybreak to collect wood for a fire to send a signal. By this time the search planes were out looking for them. Although they circled the Island, the searchers didn't see the boat because it had sunk. So Joyce was told they had drowned. The men were there till the afternoon when they spotted a cray boat. They lit their fire and the boat came over and took them safely back to Portland.

Bev Watson says she initially worried when Gary 'Ten Bins' started diving. 'Life at sea is always dangerous whether you're working a prawn trawler or whatever. There are accidents all the time.' Gary Watson did indeed die in a boating accident in 1991.

Many of the women experienced difficulties in adjusting to the lifestyle in southwest Victoria. As a Melbourne girl, Bev Watson says she tried not to stay in Portland very much. 'I hated Portland, absolutely hated it. It was the pits. It's very pretty but the weather is terrible.' Plus it's a long way from anywhere. When Bev first went there in the 1970s, the drive to Melbourne took seven or eight hours. The road surface was terrible and some sections weren't sealed and car tyres were frequently punctured.

Sandra recalls that when she and Len McCall first arrived in Port Fairy in 1968 they were living in a little shack at the back of a fisherman's house. They didn't have any children and she soon became bored living in the small town. She couldn't get a job because they only had one car and Len needed her to help drive the abalone to Melbourne.

Some of the locals were resentful of the money the divers appeared to be making. Esme Thiele recalls an incident on the wharf at Port Fairy. 'I was there while Murray was unloading his catch. Quite often there was a man who would say silly things to Murray like, what was he going to do with all his money. One day Murray turned to him and replied, "Well, you have to line your coffin with something!" The man never spoke to Murray again.'

Shirley Plummer worried about Lou every time he went out diving, until he phoned her when he had finished. Shirley was living in Geelong at the time because her children were still at school. 'He would come home at weekends if he could, but a lot of times he didn't and I would get mad because I wanted him home.' The responsibility of looking after the kids fell on her shoulders when he wasn't there. 'When my daughter had a car accident one night,' Shirley says, 'he wasn't there to help me deal with that.'

She recalls the time Lou rang her when he was hospitalised with the bends. 'I raced up to see him to make sure he was all right and there he was on oxygen and having a great time flirting with the nurses.' Shirley remembers Lou lining his wetsuit with newspaper to keep warm. 'He'd come home with newsprint and cartoons all over him and it would take a couple of days before it would wash off.'

The abalone divers were considered a wild bunch in the early days and some locals thought they were not to be trusted. Jan Braun recalls that when she first went to Portland on a camping holiday with her girlfriend, the first thing they were told was to keep away from the abalone divers and the sailors. 'We were warned off the abalone divers at first. But I met Jurgen at a party on New Year's Eve. He spoke to my girlfriend and me that night and a couple of days later he came round to where we were camping and asked me if I'd go for a drive to Port Fairy with him. So I did. And that was that. We were engaged on my 21st birthday in 1969 and we were married in May 1970.'

The last word belongs to Sandra Downes: 'In the early days we had a lot of fun and there was a lot more laughter. In later years as they got older, everyone became more interested in making money and went their separate ways. Those early days when we weren't making much money were so much better.'

11. Others reflect on the Zone

Charlie Cooper

Senior Fisheries Officer, Department of Primary Industries

Charlie Cooper's job is all about community policing. But instead of slamming everybody for every minor misdemeanour, he knows he's got to live in the community and he needs to keep a lid on the people who aren't quite conforming for whatever reason.

There's a culture in Portland among Fisheries officers that they get on well with the commercial fisherman and even with the poachers. Charlie describes one reformed poacher as 'A lovely guy. He looks as rough as guts and talks pretty rough but he's a nice bloke. His partner in crime was "Maori Pete", who died in the water in the Northern Territory. I think he was pearling at the time.

He was moving from one boat to another and fell out of the boat and drowned — that was it.'

Charlie says the poachers would take their abalone to Footscray, where they would get a heap of dope, spend a bit of time in a brothel and then come back to Portland broke. Then they'd do it all again. They used to dive at night, which made it relatively easy to escape detection and much harder for Charlie and other officers to find them. The poachers would clean a patch out at night and the licenced divers would go there in the day and grumble about not getting many abalone.

On one occasion at the Island two poachers were on their way back after a night fishing when some legitimate abalone divers turned up. 'I think it was the Port Fairy guys,'

Charlie explains. 'They rammed the poachers' boat and knocked at least one of them into the water. They were pretty frightened. The Port Fairy divers rang us and we went out on the boat from Portland. The poachers were very pleased to see us because they reckoned we'd look after them better than the divers. We seized their boat and our relationship was such that I jumped in the boat and came back to shore with them. The other divers asked if I'd be okay, because these two had a fearsome reputation. I think we forced them to go to South Australia because we hassled them so much around here.'

According to Charlie, the poachers were getting loads of about 100 abalone. Not huge volumes, but they were getting those numbers fairly regularly. He doesn't think they ever made much money out of it, but were attracted by the thrill of it all. They were lovable rogues, he says. And they were okay when they were caught, not getting aggressive, angry or upset. 'They were decent about it all and we were pretty decent with them because we knew them. But some of those Melbourne crews weren't like that at all. They were nasty.'

In the 1980s these Melbourne-based crews were well equipped with two-way

radios and fast cars. Although Charlie says they were very close to catching some of them, the poachers were scared off by the police, who were pretty ruthless, he says, and just wanted to get them out of town.

Charlie remembers one poacher who was a bit of a rogue, but a decent sort of family bloke. He recruited other young men to poach for him and then he would take the abalone and sell it in Melbourne. They caught him a couple of times. On one occasion, Portland police caught him in Narrawong, going back to Melbourne at night with a heap of abalone. Another time they found abalone on his premises, in an old wood shed near the boat ramp he rented.

Sometimes poachers left abalone in the water and then come back later to collect them. 'One time we waited in the bushes and nabbed them when they came back. We caught another local guy with his crew, who were all drug users and poached abalone to support their habit. Again they weren't nasty guys and I still see one even now. The poaching and the drugs seemed to go hand in hand. Easy money, that's what it was all about.'

Charlie was involved in chasing the notorious poacher Cam Strachan in Gippsland in the late 1970s. One time they caught him he tried to do a deal with the Fisheries officers. He said, 'Okay you've caught me, but can I keep my boat? Can I keep my gear and my wetsuit?' But they refused and took everything. Once a poacher is caught their boat, car, all their diving gear is immediately confiscated. Sometimes the inspectors had to put them on a bus to send them home. 'They lose their cars and get fined a lot of money. We had a crew who had taken 100 or 150 abalone and they were fined around $10,000 to $12,000 each for a first offence.'

Charlie joined the Department of Fisheries and Wildlife and his initial attraction was wildlife. Then in 2000 the officers had to choose to be a fisheries or a wildlife officer. Charlie chose fisheries because he is a fisherman and these days he fishes off the pier or out at sea and does a bit of snorkelling.

Clarke Espie

You've got to learn to be a survivor

Clarke Espie was one of the divers who was there at the start in the very early 1960s and worked for Frank Matthews of Marine World around 1963. Clarke dived the Western Zone around 1965 before shifting zones and believes it was Frank Matthews who began the abalone industry in Australia. 'I don't know of anyone who was exporting before Frank.'

Clarke remembers many of the key players of the early days. Phil Sawyer, he says, was a larger-than-life character and got the nickname 'JC' because he had a long bushy beard: 'They reckon he used to be able to walk across water to find an ab. He also told wonderful stories, so he was seen as almost a messiah figure. He could find fish in remote areas and come back with a big catch. He pulled in some big, big catches.'

Ricky Hale, aka 'Maori Pete', was a Tahitian Maori diver who Clarke remembers driving a red and white Oldsmobile. According to Clarke, he had the first 19-foot (5.8-m) Haines Hunter, and the boat won the Sydney-to-Newcastle ocean race. Ricky put twin 100-horsepower motors on it and it was said to be the most powerful abalone boat in the country.

Clarke remembers one day when Ricky had been working five hours non-stop in 80 feet (24 m) of water and came up with the bends. The other divers sent him down again to decompress, but they put him down to the wrong depth. After a couple of hours at around 40 feet (12 m) he became worse. At that point they raced him to a decompression chamber in Bairnsdale, about 275 km away. The doctors said he would have died in another couple of hours. Ricky was back diving in a couple of weeks.

It was unbelievably hard and physically demanding work. Clarke recalls how the techniques and equipment were primitive, with little if any industry-specific equipment available. He maintains that Frank Matthews' equipment and techniques were very different to everyone else's, especially when compared with today.

'We used to do all our diving from the beach in those days with 1000 feet [305 m] of hose. That was difficult because we had to swim or walk through the surf with a big coil of hose over our shoulder, uncoiling the hose as we went. In those days you couldn't get floating hose and so the diving hoses sunk. We would walk out from shore with potato bags and collect shellfish in these. Then we'd surface and give a "hoi" to our assistant, who'd pull us in to shore through the surf, along with our potato bags and the hose. We'd either carry the bags up the cliff face or we'd set up a flying fox to take the ab up the cliff to the high point. Once on shore, it had to be transported anything up to a kilometre from the water's edge to our old four-wheel drive, then to the factory for processing. That's how it continued for the first three or four years when Frank was operating as a processor.'

Improvements in equipment over 50 years or so has been extraordinary. Clarke says that the fishery developed its own technology because there was nothing available. It didn't take long before the divers started developing their own systems and fabricating their own equipment. Most divers had a trade background or were practical men, so they could turn their hand to most jobs and adapt or build equipment to suit their needs. 'The hookah diving with a compressor and hose on board with breathing apparatus and mouthpiece all evolved over maybe five or ten years — and as time passed it got better.' These days they have access to the most modern equipment and technology with good information, so they can operate more effectively. Clarke's son Jamie carries oxygen on board that he breathes between dives to flush out his system.

Back then they wore a weight belt, but nowadays he uses a weight vest that covers the trunk and distributes the weight evenly. This makes it less physically demanding on the diver's back. Some of the air coming through the compressors in the early days was foul with fumes coming down the hose under pressure and these created headaches for the divers. 'You just don't hear of that these days.'

Clarke reckons not everyone can be a diver. 'It might sound glamorous but it's darn tough work and often in adverse conditions. It can be rough. It can be cold and miserable. The water can be dirty. You've got to be committed, and if your heart and mind are not in the business you won't survive because it finds you out very quickly. You're alone on the bottom and it's a repetitive job. You've got to be thinking all the time and get to know the terrain. You've got to learn to be a survivor because you've got a responsibility to yourself and your deckhand to get the job done and get home safely.'

Frank Matthews

The golf connection

Frank Matthews is a quiet man who began his career as an architect, although his first love was diving. A champion spearfisherman who established the Melbourne Skin Diving Club in 1950, he is also acknowledged as having founded the Victorian abalone industry.

Frank, who never dived in the Western Zone, established the first processing plant after golfer Peter Thomson told him about the popularity of abalone in Asia. 'Peter was touring Asia and became aware of the insatiable appetite Asians have for abalone. He knew I was diving for oysters and mussels and asked whether I knew if there was abalone in Australian waters. I said I knew

where there was plenty. He began looking for a market and it took off from there.'

Matthews and Thomson decided to set the wheels in motion and begin the abalone industry. Frank decided to do more research and packed his bags and travelled along the Victorian coastline for several months to find out more about the abalone populations in the state. The information he gleaned from that first field trip gave him valuable data about the best varieties to harvest and their different environments.

Frank approached Fisheries in Melbourne but they were not interested, believing there was no market for abalone.

He wrote to experts in California, who gave him the information on canning and processing and he set up a small factory in Melbourne. He employed a food chemist, developed the first product and then had to send it to Singapore a couple of times before it was accepted for the market. Frank believes that was the first time abalone had been exported from Australia. 'We began about 1962 and once we started processing abalone then other processors like Kraft and Safcol saw there was something in it and started offering money to divers that we couldn't match with our small factory. So we started diving for other factories from then.'

Frank now lives in Tasmania.

Pat Matthey
Lucky to get away with his life

Now a retired commercial fisherman, Pat Matthey moved to Port Fairy in 1941 as a four-year-old lad and left in 1973. His father was the local cop in Port Fairy.

Pat wasn't an abalone diver but well remembers when the first wave of abalone divers swept into town in the early 1960s. He got to know those divers well. He can't remember anyone diving for abalone before then. He says that if an abalone diver got into trouble on the water, he would be phoned to go out and rescue them because he knew the waters and had a big boat.

One of the most memorable rescues was in the mid 1960s and Pat swears the men were the luckiest abalone divers ever. A gala event was being held in town to celebrate the opening of the first licenced restaurant in Port Fairy, the *Tatra on River*. Decked out in his dinner suit, Pat and his wife were about to start their meal when a police officer walked in and said an abalone boat was missing off Killarney.

Still wearing their dinner suits Pat, his brother-in-law Brian Newman and a couple of other locals launched Brian's boat and went looking for the men. On their way to Killarney Pat saw a flash out of the corner of his eye and told Brian to bring the boat around. After 15 minutes searching the area they had found nothing and the others thought Pat was seeing things. 'But there was a definite flash — not in the sky, but on the water.' Pat lined up the stars to guide them. When Brian turned the motor off, they heard voices calling and were able to rescue the men.

'I think it was Harry Bishop and his deckie. They were the luckiest blokes around. We were heading down much closer inshore and the northwesterly was blowing them out to sea. Their boat wouldn't have lasted until the morning. They said they had one match left and had dipped one of their singlets into the petrol and then lit it. If I hadn't seen that flash, well ... who knows.'

The story doesn't end there. The celebrations were still going on at the restaurant. The men hadn't had time to go home and change before the rescue and Pat recalls how it was 'damn cold in a dinner suit on the water'. When they got back to *Tatra*, 'We sat down and were about to start our main course when we were told the restaurant had to close by 11.30 pm. Everyone was up in arms saying how we'd just rescued two ab divers. The owner apologised and said he still had to close, but that we could be his guests in the kitchen. So we sat in the kitchen in the dinner suits we'd worn during the rescue and enjoyed our meal with our wives in their evening gowns.'

As in any discussion about diving and fisherman, the conversation always turns to sharks. In Port Fairy, the talk usually centres on sharks at Lady Julia Percy Island. Perhaps the most talked about and most gruesome incident at the Island was in November 1964 when Henri Bource lost his leg to a shark. Pat was on board the boat that day.

A group of divers from a Melbourne diving club had contacted Pat's father-in-law Walter because his boat was the biggest and fastest in the area. 'I said at the time that if they were going to dive at the Island that there was every chance they'd get whacked by a shark. Everyone knew there were sharks at the Island and you took your life in your hands if you dived there.

'When we arrived at the wharf on the Sunday morning to go out, Frank Newman, a fisherman, was waiting. We asked him why and he said he'd never seen anyone taken by a shark and that this might be his last opportunity. Frank was convinced he would see a shark attack that day and he was watching and waiting. He must have had a premonition. He was spot-on.'

Their boat left with the divers still keen to dive. As they got near the Island, Walter asked Pat to take the wheel. He pleaded with the men not to dive but they were determined and claimed they could handle sharks if they had to. Henri Bource's job on the day was to tick everyone on and off as they went in and out of the water, so that they had a record. Then Henri and Jill Radcliff decided to jump in and have a look around. They were only a few metres from the boat and treading water. 'His flippers must have looked like a seal and the shark just came up and took his leg — just below his knee.'

Pat radioed the shore and when they docked the local doctor Dennis Matthews was waiting. 'We had him on deck for an hour and 40 minutes. My wife was on board making tea for them. They put a tourniquet on his leg and luckily he was wearing a medal around his neck with his blood group on it. Dr Matthews had organised blood but when we arrived at the wharf and he checked Henri, he said he'd "gone". The others begged the doctor to keep trying and so he produced a needle and thrust it into Henri's chest. There was a slight tremor, then he started to get some colour.'

Henri was lucky to get away with his life that day. For those with a strong stomach, there is footage on YouTube of him on board the boat that day.

Alistair 'Hag' McDonald
Fish wholesaler in the old Safcol building

Alistair McDonald reckons Maurice Selly was the first person to fish for abalone at Portland in the early 1960s. Maurice and his deckie would drive to the wharf in an old Land Rover. They had a hookah with an air hose in the back of the vehicle and would head out from the Breakwater.

Maurice worked as a chef at Erskine House in Lorne through the summer and Alistair says, 'He'd come ab diving here in his off time.' Alistair was still at school at the time and he and his mates were often out skin diving when Maurice and his deckie were diving there. 'Maurice used to value-add by dipping the shells in hydrochloric acid to clean them up and they'd sell them to tourists.'

The abalone divers used to gather at the Henty Hotel in Portland. 'They took over the Henty,' Alistair remembers. 'There were probably 30 or 40 boats working out of here. Phil Sawyer was one of the first, also Bob Ussher, Ron O'Brien and Ricky Harris. They were all Sydney guys. There was Jurgen Braun and a French guy called Florian, Noddy Hill and Bernie Morton and Ten Bins. One guy's boat blew up one day on the wharf and burned. It was an "accident" of course, because they were mad in those days. He was probably careless when he was refuelling, but there were suggestions he did it on purpose.'

At that time there was a wharf out the front of the old Safcol building where the divers would bring the abalone in. There were always fights. 'Someone put Ronny O'Brien's head in a bin of abs. I think it was Dick Kelly. One day there was a hammer lying around and Dick was annoyed with Ron, aka "Whacker", who was some sort of kung fu or karate expert — he used to say he could kill you with one kick. He picked Ron up and dunked his head in the bin of abs. The first thing Ronny did after that was grab the hammer and say, "Come on, I'll take you on." So much for being the karate expert — he used a hammer! Things calmed down after that and they went for a beer.'

Alistair remembers another guy who bought poached abalone. They would pull the meat out of a king crab and fill it up with the abalone meat and send it to Japan that way. 'That's how king crab started to get exported. It didn't matter if they were dead or alive as long as they were filled with abalone. The authorities never caught him.'

Clark Smock
A wild industry

Clark Smock, one of the brash new breed of divers who entered the industry during the heady days of the consolidation of licences, is an enigma in the Western Zone. He is one of the five original owners of Sou'west Seafoods and a major shareholder, yet he's not a director. He came close to buying a licence in the Zone but never held a full licence.

Clark was a dive instructor who also played water polo and swam. He heard Gary 'Ten Bins' Watson's abalone licence

the fishery believed they were entitled to a golden handshake when they sold their licences. So they put forward the two-to-one scheme to Fisheries. That's when a lot of the older divers sold their licences. As Clark explains, 'The rationale was that this would reduce the impact on the fishery, but what they didn't plan on was that the new guys were young and eager to go. The guys who they bought out were probably fishing 35 tonne between them. In my first year I fished nearly 60 tonne. It was a completely different attitude at this time.'

Clark remembers Dick Cullenward as one of the true characters of the industry who had the physiology to deal with, and push, things to the limit. Dick virtually 'owned' Lady Julia Percy Island because he dived there all the time. He says Dick would see maybe six sharks a year and got to know their personalities. One time he was working off his new boat *New Toy* and, when he went to get back into the water after a break, a big shark swam through the middle of the two hulls. 'It was about to exit out of the aft end when Dick stepped off the back and landed on it. He said he got wet to about his knees and the shark blew him back into the boat.'

was for sale at the time of the two-to-one consolidation of licences. 'I put word out that I wanted to buy a licence and Ten Bins said he had a Western Zone one for sale. So I put a down payment on it then a guy in the Central Zone had one and I put a down payment on that one as well. Now I had options on licences in two zones that I couldn't consolidate. Rod Crowther was in the same situation and came to me and said he'd buy mine in the Western Zone. I agreed, provided I could buy his in the Central Zone.'

At the time he purchased his original licences, Clark says the government was trying to diminish the impact on the fishery. The original divers who had established

Clark remembers Rod Crowther taking delivery of a 7-metre Cougar Cat. The boat's designer delivered the craft and declared it was 'Flaming unsinkable — you can't sink these boats.' This statement was like a red rag to a bull and the six divers who were around at time smiled at each other, 'Really?' they said. The conditions that day were miserable, with some 'nasty stuff' going on. 'We thought we'd test his theory. We went out and whoever was at the wheel throttled down the face of a 2.5-metre wave and then backed off. Everyone yelled #@!* and the boat popped straight up. Then one of the guys ran up a wave sideways and then throttled off. The whole boat was sitting on top of the wave, then it slide sideways in white water and froth. The props had nothing to grab on to and were screaming. The boat was sliding sideways down the wave and rolling towards the rock face. We were all holding on — waiting. We kept doing these sorts of things and the owner was yelling, "You guys are crazy. Let me off." We didn't sink it even though we tried our very best.'

Clark remembers days when, after spending ten and a half hours underwater, he'd get out, sit down and just go, 'Blahhh' because he couldn't talk. Jurgen Braun's work practice was to work four or five hours and then get out, no matter whether it was a perfect day or not. 'That was Yogi. Dick Cullenward would go out to the Island while Bob Ussher would go deep at Bridgewater and the Capes. Phil Sawyer would beach-launch at Bridgewater and work the cliff faces. Ten Bins would just say he'd caught ten bins, whether it was accurate or not — it didn't matter.'

Clark sums up the industry nicely. 'When you've had a full day and you get on board and you are physically beat up and your deck is full of fish and you round a corner to the calmer parts of the harbour, there's a mixture of pride and accomplishment. It's a wild industry.'

Tassie Warn

Why catch six fish when you can catch four?

Typical of the old-school abalone diver, Tassie Warn only spent about nine months in the Western Zone diving at Portland. He is a fourth-generation fisherman who's only ever been a diver. He can remember his uncle diving for abalone in Bass Strait in the late 1940s. He's also a crusty, colourful, no-

nonsense kind of man who saw many changes in the industry before he left around 2006.

Tassie arrived in the Western Zone just before introduction of the Zones, which locked many divers out for good. 'A group of us were sick of working Mallacoota and Eden and those isolated places. In 1966–67 there were plenty of abs and we enjoyed Portland, but the weather was cold.'

He has many great memories of the Western Zone pioneers, but Tassie is quick to put those early abalone divers into two camps: adventurer/explorers and the money hungry. Because of this, he says tensions flared from time to time between the two camps. 'For a lot of us it was all about finding new reefs. We would spend a lot of time doing this while the others never did a bloody thing. They just waited until someone found a good patch, then they'd come in on it. These were well-known and well-liked blokes in the industry, but in the early days I just used to think they were arseholes. They'd never go looking.'

In the 1960s the men worked and played hard, and Tassie was one of those who would move up to the southwest coast of New South Wales when the weather got bad.

'Most of the early divers were in it for the fantastic lifestyle and good income. We'd work all day and load up with grog and drive to Melbourne, then turn around and do the same thing the next day. We all drank too much, but it was bloody hard work with long hours.'

The wild ways and live-for-today attitude of the divers caused a few rumblings around the sleepy towns where the men lived. 'If it was a lovely sunny day and I was at the pub or just messing around, people used to ask me why I wasn't working. I'd say to them that it was too nice to work — and I wasn't joking.

Even though it was a lovely day, if you knew anything about the ocean and if there was a roll up or an onshore wind, then it was no good for diving. I just got sick of telling them this.'

The divers made the most of the lifestyle and money they earned, but they were always aware of the dangers, particularly the bends. Tassie remembers some of the divers being better than others at keeping track of the time they spent underwater and then correlating these with depth charts. 'We didn't consider the risks back then because it didn't often happen. And because it didn't happen, we didn't believe it when the experts told us the risks. I remember when a US dive medicine expert arrived around 1966. He gave a talk to us and started off by saying, 'Well, I'm looking at a room full of dead men,' which was a fair comment. I got bent a couple of times in the leg, which caused me a bit of pain for that day. However, while a lot of us would spend six to eight hours in the water, most of my work was under 10 metres.'

These were men who earned big money and who could spend it fast. Tassie tells a yarn about a couple of the Portland divers. 'They rocked up in the latest model Holden pulling a Haines Hunter, which was the boat of choice in those days. They roared up and backed it straight into the water at about 30 kph. The next thing the car's in the water and flooded, and the driver's just sitting there grinning. Within 18 hours he'd dried and cleaned it, driven to Melbourne, traded it in and came back with a new Falcon GTHO.'

Tassie says he's been darn lucky to be involved in the industry and that it provided him with the lifestyle he liked. 'If I needed money I would go out to work and get it. Then I'd spend it and even got into debt. I'd then think, "Shit, I'd better get out and go to work," and I would. It worked well and I was happy. It was a good system for me and my wife, but not everybody thinks like that.'

While the industry has been good to Tassie, the hype about all abalone divers being rich annoys him. According to him, sure, there were those who saw it as a way to make big money. 'They got into the industry, made their money and moved on. Some are now very wealthy. Then there are the other happy-go-lucky, lifestyle divers who are financially comfortable but they never exploited the chances they had. I'm definitely one of those. Anyone who had half a business brain should have been a multimillionaire — but most of us weren't.'

After a lifetime in the industry, Tassie remains passionate about protecting the fishery for the future. He believes that if you leave a surplus of breeding stock, you can't go wrong. 'Despite the increase in live fishing, I would have closed the season in the peak breeding season. But because the warm water moves along the coast, you couldn't have one blanket time when you'd take, say, two months out of the season. I didn't see that as being such a hard thing to do, but others didn't see it that way. I'd say to them to catch decent-size fish and not the young ones. Why catch six fish when you can catch four? Most didn't agree with me.'

Tassie retired a few years ago and continues to live on the Bellarine Peninsula in Victoria.

Frank Zeigler
They know the risks
Frank Zeigler is an ex-police officer who's been involved in search and rescue as well as routine police work. He now has a dive business in Portland. Frank came to Portland in 1979 with the police force. At that time there was very little infrastructure for diving in Portland. He set up his business, Professional Diving Services, while he was still in the force.

At that time the only abalone divers in Portland were Gary Watson, Dick Kelly, Phil Sawyer, Derek Fieguth, Bernie Morton, Rick Harris and Bob Ussher. Frank remembers them as 'very colourful characters, and all very much pioneers of the time. Dick Kelly, for example, bought his general fishing licence for a shilling. He showed me a copy of his original licence at one stage. I thought

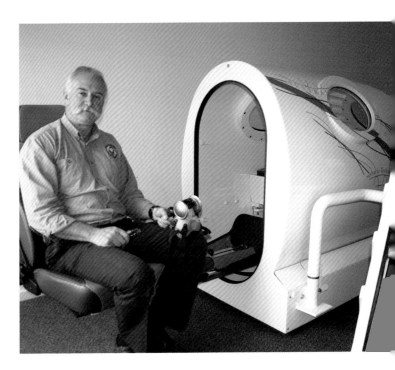

that was amazing. He grumbled when a couple of years later it went up to £1 [$2].'

When Frank arrived in Portland, the abalone divers had been there for almost 20 years. They were still doing things the way they always had done things. Frank remembers Dick Kelly coming in one day and saying he had been diving deep that day — about 14 feet (5 m). A week later Frank went out for a dive with him with a depth gauge. Dick was actually diving in 18 to 20 metres of water, when he thought he was in 5 metres. He was pushing the known limits of dive tables. 'In those days they weren't using Australian or Canadian tables. They weren't using any tables.'

Frank was amazed at the number of divers who were getting the bends on a regular basis. He set up the diving business and often worked until the early hours putting a number of divers on oxygen and trying to ease their symptoms. 'They should have called *HMAS Penguin*, which had a hyperbaric chamber where they could be treated. They couldn't or wouldn't because the weather report was too good for the next day and they'd have to be back out there again.'

Daniel Ussher offered to trial one of Frank's new 'dry' suits: 'We were trying to get the guys into dry suits, because thermally they are better than a wetsuit. It keeps you dry underwater because it's fully sealed, there's a neck seal and wrist seal, so no water gets in. We provided Daniel with a brand new dry suit and out he went. When he brought it back in it was wet through. When asked if it had leaked, he said, "No, I forgot I was in a dry suit and had a piss in it!" So we adapted his suit with a "pee" zip, plus other innovations.'

Frank remembers a tragic incident involving Dick Kelly's deckie. They were working in extreme conditions. The deckie went over the bow of the boat. The boat was then pushed towards the cliff. Dick had no option but to put the boat into reverse and he took off the head of his deckie. 'It was fatal. He couldn't do anything else. That devastated Dick, absolutely devastated him. Dick was a man of means and I know he looked after the family. The coroner's finding was death by misadventure.'

Frank thinks some of the divers are seen as wealthy and eccentric, but he says they are also good community members. He says they put a lot of money into research and development of the abalone industry as well as a lot of money into the town — and the local hotels. Bob Ussher was a good example. 'He'd be embarrassed to tell you about the things he's done for the community,' Frank says. 'He's a great guy but he put himself through so much pressure. The name of his company, 'Ard Gowan, says it all, because in the early days it was hard going. He was a larrikin. He'd drive his Rolls Royce convertible down the main streets of Portland waving at the locals. When I'd only just arrived in town, we got a call to a medical emergency at his house. His wife was saying, "Bob's not breathing". We did CPR on him and he came back, then we put him in the ambulance. The reward was to later see that smiling face as he waved at us from his Rolls Royce convertible.'

To tackle the poaching problem, the divers got together and decided they wanted Fisheries to assist with enforcement. They had a couple of the biggest prosecutions in Portland with the highest penalties for poaching. One Portland local was convicted of trafficking in abalone and poaching. 'I think he was fined $8 per fish, which was the highest penalty paid. That was quite significant.'

Frank explains the poachers were pretty smart and well organised with buyers for their catch. Some of the poachers were the go-betweens, so they never actually got involved and they would be paid partly in cash and partly in marijuana. Marijuana was used as effective pain relief for decompression injuries. The divers would use any form of analgesic for pain relief. Gary Watson used Bex powders and was always asking his long-suffering wife Bev to get him a Bex and a cup of tea.

As a member of the police force, Frank admits that it was hard to be impartial because he knew the divers so well. The senior sergeant would say, 'Another of your bloody ab divers has got a problem, Zeigler. Go and sort it out.' And he usually did.

12. Glossary of terms

Access licence holder: Holder of an access licence issued under section 38 of the Fisheries Act 1995.

AFAL: Abalone Fishery access licences

AQIS: Australian Quarantine and Inspection Service

AQMS: Abalone Quota Management System

Awab: Means abalone (from Spanish abulón)

Bends or being bent: Bone necrosis, a painful, sometimes fatal illness in which nitrogen bubbles form in the diver's bone marrow and escape through the bone joints. It can develop during or after diving. Any organ or tissue may be involved and its presentation can vary from the acute to the chronic.

Bitey: A shark

Bombora or bombie: Waves breaking over a submerged rock shelf.

Bottom time (BT): The total elapsed time from when the diver leaves the surface to the time that the diver begins the ascent, measured in minutes (nearest whole minute).

Breathing tubes: Tubes or hoses attached to a regulator that are designed to supply air or gas to the diver, carry away expired gas and operate at near ambient pressure

Compression chamber (recompression chamber): Surface chamber in which person may be subjected to pressures equivalent to or greater than those experienced while underwater, or under conditions, which simulate those, experienced on an actual dive. Used by divers who have the bends.

Deckhand, deckie (aka 'sheller'): Person responsible for the supervision of the boat, the diver and collection, measurement and containment of abalone catch. This includes the use of all safety equipment, supporting, recovering and attending the diver, and initiating and operating all safety procedures.

Decompression sickness: See Bends or being bent.

Dive plan: Procedure by which any additional precautions are implemented for a particular diving operation.

Diver days: Days per month a diver dives

Diver's hose: A single length of an approved hose for breathing gas, which carries gas from the surface to the diver.

Dive site: Underwater location where work is performed. Also any surface zone used to tend or supervise the diver.

DPI: Department of Primary Industry

Haines Hunter: Often the boat of choice among abalone divers

ITQ: Individual transferable quota

Limiting line: A line shown in decompression tables that indicates time limits (bottom times) beyond which decompression schedules or exposures are deemed less safe. Diving for periods indicated below the limiting line carries a greater risk of decompression sickness, and this risk increases with time increase. See Bends or being bent.

LML: Legal minimum length

MAFRI: Marine and Freshwater Resources Institute

Marine Safety Victoria: The statutory body responsible for the enforcement of the Marine Act and the survey and approval of commercial vessels and water craft in the state of Victoria.

MOU: Memorandum of Understanding

MSC: Marine Stewardship Council

Nitrox: A mixture of oxygen gas and nitrogen gas, together which contain no less than 21% oxygen.

Quick release: The ability to be immediately released from the secured position by the single operation of one hand.

Scuba: Self-contained underwater breathing apparatus

Shark POD (protective oceanic device): A device that surrounds the diver with an electrical field to repel sharks. As a shark approaches the POD, the field first causes discomfort and then muscle spasms, which will cause the shark to flee. It affects receptors in the shark's snout and its central nervous system.

Sheller: The term previously used for deckhand, who also had the added duty of shelling the abalone at sea.

Shuck: To remove the shell from an abalone, done by hand

SIV: Seafood Industry Victoria

SPC: South Pacific Canneries Pty Ltd, Collingwood, Melbourne

Surface interval: The time a diver has spent on the surface following a dive, beginning when the diver surfaces and ending as soon as the diver commences the next descent.

TAC: Total allowable catch

Therapeutic recompression tables: Tables used for the treatment of decompression sickness and other pressure-related injuries. See Bends or being bent.

VADA: Victorian Abalone Divers Association

VSL: Voluntary size limits

WADA: Western Abalone Divers Association

13. Photo captions and credits

p vi: Peter Riddle. *Photo*: Rob Torelli.

p 8: Early licences showing how the cost skyrocketed from $4 to $200 between 1967 and 1969. *Photo source*: Murray Thiele.

p 9: 1978 media coverage. 1967 Abalone fishing requirements. *Photo source*: Murray Thiele

p 10: The abalone 'fleet' at Propeller Bay. *Photo source*: David Forbes.

p 11: Glenn Plummer's abalone boat.

p 12: Len McCall's letter advising of the establishment of two Victorian zones. *Photo source*: Len McCall.

p 13: Scielex abalone-shell measuring board.

p 15: Virus-affected abalone next to healthy abalone. *Photo source*: Rob Torelli.

p 17: Top to bottom, Rob Torelli at Lady Julia Percy Island. *Photo source*: Rob Torelli. Catch in David Fenton's boat. *Photo source*: David Fenton.

p 19: Hyperbaric chamber at Frank Zeigler's Professional Diving Services.

p 20: 1967 invoice from South Pacific Canneries Pty Ltd. *Photo source*: Murray Thiele. Letter to Sandra McCall advising of her share holding. *Photo source*: Len McCall.

p 33: Boat laden with abalone. *Photo source*: David Fenton.

p 35: Sou'west Seafoods Co-op on Port Fairy Wharf. *Photo source*: Len McCall.

p 48: Abalone bed. *Photo source*: Rob Torelli.

p 50: Dr Jeremy Prince. *Photo source*: Jeremy Prince.

p 51: WADA workshop, 3 October 2007.

p 52: Rob Torelli about to start work at The Crags.

p 53: Peter Riddle doing a bag changeover. *Photo source*: Rob Torelli.

p 58: Early abalone divers campsite. *Photo source*: Tony Jones. Bob Ussher launching his boat. *Photo source*: Bob Ussher.

p68: Crags Beach coast with Lady Julia Percy Island seen on the horizon.

p 82: Newspaper coverage of Harry Bishop being stranded at Killarney.

p 88: Dick Cullenward's boat *The Double Uggly*.

p 91: Derek Feiguth. *Photo source*: Derek Feiguth.

p 101: Victor O'Brien on Ron O'Brien's boat at Lady Julia Percy Island. *Photo source*: Jamie Espie.

p 103: Gary 'Ten Bins' Watson (right) and his deckie Peter Doran. *Photo source*: Peter Doran.

p110: Phil Sawyer drives the hydraulic bag carrier. *Photo source*: Konrad Beinssen.

p 123: Victor O'Brien at Lady Julie Percy Island. *Photo source*: Jamie Espie.

p 130: Fox Statistics. *Photo source*: Craig Fox

p 135: John Hollingworth. Kids helping shell abalone. *Photo source*: John Hollingworth.

p 143: Harry Bishop and Bernie Morton in his Geelong football jumper to ward off sharks. Bernie Morton (centre) and Harry Bishop (right). *Photo source*: Noel Middlecoat.

p 145: Catch on Peter Riddle's boat.

p 154: Rob Torelli searching for abalone through a seaweed bed. Rob Torelli at Lady Julia Percy Island. Peter Riddle at work. *Photo source*: Rob Torelli.

p 164: Abalone divers and deckies enjoying a drink. *Photo source*: Len McCall.

p 174: Bob Hope at Ummm and Ahhh Point.

p 175: John Jehu packing abalone. *Photo source*: Murray Thiele.

p 178: Robert Rendall and Peter Riddle. *Photo source*: Rob Torelli.

p 188: Sou'west factory in Awabi Court, Port Fairy. Workers at the Sou'west factory.

p 189: The Co-op on the Port Fairy Wharf. *Photo source*: Len McCall.

p190: Sandra McCall, Esme Thiele, Natalie Cullenward preparing product. Esme Thiele and Sandra McCall selling product. *Photo source*: Len McCall

p 193: Loading abalone. Media coverage of Valmai and Robert Coffey packing abalone. High jinks to promote the opening of the Co-op. *Photo source*: Len McCall.

p 194: Sou'west Board, 2011.

p 196: Esme Thiele, Sandra McCall and Valmai Coffey preparing product to sell. *Photo source*: Len McCall. 'Sea' of abalone. *Photo source*: David Forbes.

p 198: Sandra McCall and Valmai Coffey working at the Co-op shop. Inside the Co-op shop on Port Fairy Wharf. *Photo source*: Len McCall.

p 206: Maurice Dalton and Simon McCall (diver and deckhand for Maurice Dalton respectively), Len McCall, Harry Peeters, Dallas D'Silva (former abalone manager for DPI). *Photo source*: Harry Peteers. Frank Matthews and Jim Levin carting their catch at Cape Otway. *Photo source*: Frank Matthews.

p 208: Frank Matthews with his catch. *Photo source*: Frank Matthews.

p 213: Behind Clark Smock is the mouth of the Hopkins River, which is the dividing line between the Eastern and Western Zone.

p 217: Frank Zeigler with a hyperbaric chamber.

p 220: Seals playing at The Crags. Photo source: *Rob Torelli*. The Crags.

p 225: Rick Harris and deckhand.

p 226: Living conditions were tough for the men. Tony Jones at front. *Photo source*: Bob Ussher Frank Matthews with catch